Global Forest Fragmentation

Global Forest Fragmentation

Edited by

Chris J. Kettle

Department of Environmental System Science,
ETH Zurich, Switzerland

and

Lian Pin Koh

University of Adelaide, Australia

www.cabi.org

CABI is a trading name of CAB International

CABI
Nosworthy Way
Wallingford
Oxfordshire OX10 8DE
UK

Tel: +44 (0)1491 832111
Fax: +44 (0)1491 833508
E-mail: info@cabi.org
Website: www.cabi.org

CABI
38 Chauncy Street
Suite 1002
Boston, MA 02111
USA

Tel: +1 800 552 3083 (toll free)
E-mail: cabi-nao@cabi.org

A catalogue record for this book is available from the British Library, London, UK.

Library of Congress Cataloging-in-Publication Data

Global forest fragmentation/edited by Chris J Kettle and Lian Pin Koh, Department of Environmental System Science, ETH Zurich, Switzerland.
pages cm
Includes bibliographical references and index.
ISBN 978-1-78064-203-1 (hb : alk. paper) -- ISBN 978-1-78064-497-4 (pb : alk. paper) 1. Fragmented landscapes. 2. Forest ecology. I. Kettle, Chris J. II. Koh, Lian Pin, 1976-

QH541.15.F73G56 2014
577.27--dc23

2014020717

ISBN-13: 978 1 78064 497 4 (Pbk)
978 1 78064 203 1 (Hbk)

Commissioning editor: Vicki Bonham
Editorial assistant: Emma McCann
Production editor: Lauren Povey

Typeset by SPi, Pondicherry, India.
Printed and bound in the UK by CPI Group (UK) Ltd, Croydon, CR0 4YY.

Contents

Contributors

Andrew D. Barnes, Systemic Conservation Biology, J.F. Blumenbach Institute of Zoology and Anthropology, University of Göttingen, Berliner Strasse 28, 37073 Göttingen, Germany. E-mail: abarnes@gwdg.de

Keith Barney, Crawford School of Public Policy, Australian National University, Canberra, Australia. E-mail: keith.barney@anu.edu.au

Christopher Barr, Woods & Wayside International, Hopewell, New Jersey, USA. E-mail: cbarr@woods-wayside.org

Jürgen Bauhus, Faculty of Environment and Natural Resources, University of Freiburg, Germany. E-mail: juergen.bauhus@waldbau.uni-freiburg.de

Shonil Bhagwat, Department of Geography, The Open University, Walton Hall, Milton Keynes, MK7 6AA, UK; School of Geography and the Environment, University of Oxford, South Parks Road, Oxford OX1 3QY, UK; Linacre College, Oxford, OX1 3JA, UK. E-mail: shonil.bhagwat@open.ac.uk

Virginie Boreux, Institute of Ecology, Leuphana University of Lüneburg, Germany; Faculty of Environment and Natural Resources, University of Freiburg, Germany, E-mail: virginie.boreux@nature.uni-freiburg.de

M. Jahi Chappell, Agroecology and Agriculture Policy, Institute for Agriculture and Trade Policy, Minneapolis, Minnesota, USA; School of the Environment, Washington State University Vancouver, Vancouver, Washington, USA. E-mail: m.jahi.chappell@vancouver.wsu.edu

Richard T. Corlett, Xishuangbanna Tropical Botanical Garden, Chinese Academy of Sciences, Menglun, Mengla, Yunnan 66630, PR China. E-mail: corlett@xtbg.org.cn

Katharine J.M. Dickinson, Department of Botany, University of Otago, PO Box 56, Dunedin 9054, New Zealand. E-mail: kath.dickinson@otago.ac.nz

Raphael K. Didham, School of Animal Biology, The University of Western Australia, 35 Stirling Highway, Crawley WA 6009, Australia; CSIRO Ecosystem Sciences, Centre for Environment and Life Sciences, Underwood Ave, Perth, WA 6014, Australia. E-mail: raphael.didham@uwa.edu.au

Timm F. Döbert, School of Animal Biology, The University of Western Australia, 35 Stirling Highway, Crawley WA 6009, Australia; CSIRO Ecosystem Sciences, Centre for Environment and Life Sciences, Underwood Ave, Perth, WA 6014, Australia. E-mail: dobert01@student.uwa.edu.au

Laurène Feintrenie, Goods and Services of Tropical Forest Ecosystems, Centre de Coopération Internationale en Recherche Agronomique pour le Développement (CIRAD), F-34398 Montpellier, France. E-mail: laurene.feintrenie@cirad.fr

Aline Finger, Royal Botanic Garden Edinburgh, Genetics and Conservation, Edinburgh, EH3 5LR, UK. E-mail: a.finger@rbge.ac.uk

Joern Fischer, Institute of Ecology, Leuphana University of Lüneburg, Germany. E-mail: joern.fischer@uni.leuphana.de

William A. Foster, Department of Zoology, University of Cambridge, Cambridge, CB2 3EJ, UK. E-mail: waf1@hermes.cam.ac.uk

Claude A. Garcia, Department of Environmental System Science, ETH Zurich, Switzerland; Goods and Services of Tropical Forest Ecosystems, Centre de Coopération Internationale en Recherche Agronomique pour le Développement (CIRAD), F-34398 Montpellier, France. E-mail: claude.garcia@usys.ethz.ch

Jan C. Habel, Technische Universität München, Department of Ecology and Ecosystem Management Center for Food and Life Sciences 85354 Freising, Germany. E-mail: janchristianhabel@gmx.de

Rhett D. Harrison, World Agroforestry Centre East Asia Node Kunming Institute of Botany, Kunming 650201, PR China. E-mail: r.harrison@cgiar.org

Chris J. Kettle, ETH Zurich, Institute of Terrestrial Ecosystems, Ecosystem Management, CH-8092 Zurich, Switzerland. E-mail: chris.kettle@env.ethz.ch

Alexandra-Maria Klein, Institute of Ecology, Leuphana University of Lüneburg, Germany; Faculty of Environment and Natural Resources, University of Freiburg, Germany. E-mail: alexandra.klein@nature.uni-freiburg.de

Lian Pin Koh, School of Earth and Environmental Sciences, and the Environment Institute, University of Adelaide, Adelaide, SA 5005, Australia. E-mail: lianpinkoh@gmail.com

Sarah A. Laird, People and Plants International, Bristol, VT 05443, USA. E-mail: sarahlaird@aol.com

William F. Laurance, Centre for Tropical Environmental and Sustainability Science & School of Marine and Tropical Biology, James Cook University, Cairns, Queensland, Australia. E-mail: bill.laurance@jcu.edu.au

Yangchen Lin, Department of Zoology, University of Cambridge, Cambridge, CB2 3EJ, UK. E-mail: linyangchen@gmail.com

Sarah H. Luke, Department of Zoology, University of Cambridge, Cambridge, CB2 3EJ, UK. E-mail: sarah.h.luke@gmail.com

Kelvin S.-H. Peh, Department of Zoology, University of Cambridge, Cambridge, CB2 3EJ, UK; Institute for Life Sciences, University of Southampton, Southampton, SO17 1BJ, UK. E-mail: kelvin.peh@gmail.com

Stacy M. Philpott, Environmental Studies Department, University of California, Santa Cruz, California, USA. E-mail: sphilpot@ucsc.edu

Ute Radespiel, University of Veterinary Medicine Hanover, Institute of Zoology, D-30559 Hanover, Germany. E-mail: ute.radespiel@tiho-hannover.de

Edgar C. Turner, Department of Zoology, University of Cambridge, Cambridge, CB2 3EJ, UK. E-mail: ect23@cam.ac.uk

Bruce L. Webber, CSIRO Ecosystem Sciences, Centre for Environment and Life Sciences, Underwood Ave, Perth, WA 6014, Australia; School of Plant Biology, The University of Western Australia, 35 Stirling Highway, Crawley, WA 6009, Australia. E-mail: bruce.webber@csiro.au

Preface

Forest fragmentation is sometimes deceptive because a forest appears to be intact, and human impact seems minimal. If there are lots of trees everywhere, what is the problem? Even remote sensing using satellites to survey larger areas often suggests that a landscape is largely covered in trees, and the area is considered forest. Yet forest fragmentation can fundamentally alter the ecology of a forested region even when roads, agricultural fields, logging or other human activities only eliminate a small proportion of the habitat. For example, forest fragmentation allows greater access to hunters who remove large vertebrates, with a subsequent loss of seed dispersal for the plants, predator control of rodents and other ecological services. Villagers and other local people can more easily enter forest fragments and remove edible, medicinal and ornamental plants, and often firewood as well. Invasive plants and animals migrate along the forest edge and penetrate into the forest, in the process fundamentally changing the biological community. Microsite changes, including lower humidity, higher wind speeds and increased incidence of fire, can all kill trees and animals at the fragment edge and alter conditions hundreds of metres into the forest. These forces often act synergistically to degrade the biodiversity of forest fragments.

Even though we understand the pernicious effects of forest fragmentation, the remaining areas of forest in the world continue to be fragmented by human activities. An increasingly extensive road network is penetrating remote areas of the world. The high prices for commodities create greater incentive to log forests for their timber and to convert natural forest into plantations of crops such as coffee, oil palm and rubber. The result is ever more forest fragmentation and the creation of smaller fragment sizes. Although the decline in traditional rural populations in many areas of the world results in less pressure for subsistence agriculture, the globalization of the world's economy and the increasing human population put greater pressure on the land for intensive agriculture, harvesting timber and the growing of crops for international markets.

In this series of 10 chapters (following an Introduction) an international group of authors cover the major questions of forest fragmentation as general questions affecting the world's forest biodiversity, ecosystem function and ecosystem services. These are not merely case studies from particular places in the world, confirming what we already know. Rather, these are carefully selected topics that provide a broad overview of the major questions confronting forest ecologists, conservation biologists and environmental policy makers. Among the key themes throughout the book are whether land sharing or land intensification represents a better way to protect biodiversity, if subsidies from carbon sequestration programmes such as REDD (reducing emissions from deforestation and forest degradation) can provide a permanent solution to forest conservation and how forest fragments will be able to persist during a changing climate. An additional topic is the extent to which government policies can reverse the effects of forest fragmentation and rebuild the forest systems that can protect biodiversity. As such, this book will make excellent reading for graduate students seeking to understand the general topic of forest fragmentation and to discern which are the major questions of current concern. Furthermore, this will provide a valuable reference for environmental policy makers, while conservation biologists and ecologists can use these essays to develop a cutting-edge research agenda.

Professor Richard B. Primack

Acknowledgements

We would like to extend our gratitude to all our colleagues at ETH Zurich, particularly within the Ecosystem Management Group, who provided encouragement and advice, especially Jaboury Ghazoul. Special thanks to Vicki Bonham and the staff at CAB International for their patience during the development of the book and continued editorial support during its completion. Not least, we thank all the chapter contributors for their reliability and responsiveness to e-mails and requests. We would also like to acknowledge David Burslem for providing additional encouragement in the initial stages of developing the ideas for this book. Finally, special thanks are due to family Kettle-Martin for tolerating the additional workload that this book generated. Perhaps one day this book might be of interest to them, but we hope the threats to forests and the rate of forest fragmentation will have reduced by then.

1 Global Forest Fragmentation: Introduction

Chris J. Kettle[1]* and Lian Pin Koh[2]
[1]*ETH Zurich, Switzerland;* [2]*University of Adelaide, Australia*

Forest fragmentation has profound global impacts. Not only does this impact on the Earth's potential to sustain biological diversity, but ecosystem function and numerous ecosystem services including climate change mitigation, food security and livelihoods are also all altered. Forest fragmentation is an inevitable result of population growth and economic development, which thus presents a major environmental challenge. This challenge is to ensure that future anthropogenic landscapes have the capacity to sustain biodiversity and ecosystem function and provision, the wide range of services upon which humanity depends. This book provides a cutting-edge review of the multidisciplinary sciences related to the study of forest fragmentation. Specifically, it addresses cross-cutting themes from ecological and social sciences perspectives to provide a valuable resource to students, forest scientists, resource managers, agricultural extension workers and policy makers. The ultimate goal of this book is to help form the scientific basis for future forest-landscape management and planning to meet global environmental and societal needs. Although the book covers forest fragmentation from a global perspective, there is some inevitable bias towards the regions of the world where forest fragmentation is most pressing and where the consequences are the furthest reaching.

We begin with a historical perspective of forest fragmentation from 33 locations around the world. In Chapter 2, Shonil Bhagwat pieces together snippets of information from case studies to reconstruct a timeline of global forest fragmentation during the Holocene (the last 11,500 years). He reveals that the timeline, scale and extent of deforestation are highly variable. In Africa, for example, genetic studies have suggested that the first occurrence of forest opening in Madagascar appears to be pre-human, possibly implying the role of environmental drivers of prehistoric land-cover change, whereas in the Americas, deforestation began during the Little Ice Age (700–1000 years before present), particularly on the Yucatan Peninsula by the Mayan civilization. Subsequently, the arrival of the Spanish and climatic change exacerbated forest loss in that region. It was a different story in Asia, where human impact on forests began earlier than in the Americas. For example, analysis of fossil pollen data has suggested that large-scale deforestation of evergreen oak forests happened between 5300 and 4500 years before present in the

*chris.kettle@env.ethz.ch

middle Yangtze basin in north China. Bhagwat concluded that forest loss and fragmentation have been major processes in landscape development in Europe during the last 11,500 years; the same drivers of deforestation in Europe are also evident on other continents, including the Americas, Australasia and Eurasia, although the timing of deforestation might be different, and the eastern Mediterranean region has some of the earliest reported dates of deforestation.

Chapter 3 fast forwards the story to the present time, as William Laurance discusses several contemporary drivers of habitat fragmentation worldwide. In recent decades, drivers of land-cover change have shifted dramatically from small-scale farming and rural development to industrial activities including large-scale farming, ranching and tree plantations. Much of this change is driven by globalization to meet rising demands for goods moving from developing to developed nations. The opening up of forests often also has severe knock-on effects. For example, logging, mining, fossil-fuel extraction and infrastructure projects are creating an economic impetus for road building in many frontier areas, which also promotes habitat loss. This often results in a vicious cycle of land development until all forests are almost completely destroyed, degraded and fragmented. In general, old-growth forests are declining rapidly across the planet and are being replaced by fragmented, secondary and logged forests and plantation monocultures. Although forest regeneration may partially mitigate habitat fragmentation in some regions, rapidly expanding plantations of non-native tree species typically have limited benefits for biodiversity.

In Chapter 4, Döbert and colleagues take on the hugely difficult and often contentious issue of conserving species in human-dominated landscapes. Conservation management is only just beginning to come to grips with the challenges faced in moving away from the traditional patch-focused approach of conservation in gazetted nature reserves towards a landscape-focused approach of conserving biodiversity in a mosaic that includes managed and semi-natural habitats. More importantly, the processes operating in the land-use matrix surrounding forest remnants drive many of the negative impacts of land-use change on remnant forest but at the same time play a vital role in long-term biodiversity conservation in fragmented landscapes. Therefore, the authors argue that the maintenance of biodiversity will depend as much, if not more, on the extent, magnitude and spatial structuring of landscape processes within the degraded matrix surrounding primary forest remnants.

In Chapter 5, Finger and colleagues review the broad range of genetic consequences faced by both animal and plant species as a consequence of forest fragmentation. The negative genetic consequences and their underlying drivers are often cryptic and notoriously difficult to identify. Not only because they involve processes that are hard to observe in the wild, such as dispersal, gene flow, mating systems and inbreeding but also because these can operate over complex spatial and temporal scales. This chapter identifies many of the key ways in which molecular approaches can be applied to advance our understanding of these critical processes and how species traits might influence their vulnerability to forest fragmentation genetics. Some species appear to be far less vulnerable to fragmentation genetics, and this chapter asks the question of why this is so and what traits are important in the context of forest species. By reviewing the state of knowledge and critical challenges, the chapter provides a stage upon which to set priority areas for future research.

How would climate change interact with human drivers of land-cover change to influence the future trajectory of forest loss and fragmentation? And, perhaps more crucially, how would forest loss feedback to and influence future climate change? These are essentially the questions dealt with by Richard Corlett in Chapter 6. Forest fragments are vulnerable to climate change, both directly and through changes in fire frequency and intensity. Gaps between fragments will also reduce the ability of forest species to track climate change by movements along climate gradients. Conversely, forest fragments retain carbon that would otherwise be lost to the atmosphere and contribute to global warming.

Therefore, protecting and restoring fragments can help mitigate climate change by reducing emissions and increasing sinks, especially if larger forest areas are unavailable. If fragments are actively restored, it makes sense to use plants adapted to the expected future climate, even if this involves the 'managed translocation' of species outside their current ranges.

One of the most contentious issues in conservation science surrounds the debate on whether biodiversity should be integrated on the same land (land sharing) or separated from agriculture (land sparing). In Chapter 7, Klein and colleagues discuss the virtues of both approaches using actual case studies. They argue that, with regard to land sharing, there is no simple trade-off between biodiversity and yield, and the complexity of crop yield–biodiversity relationships has not yet been fully investigated. Land-sparing management has to be considered at different spatial scales, as agriculture in large-scale land sparing generally does not benefit much from ecosystem services mediated by forest biodiversity, although small forest remnants can enhance biodiversity in large agricultural and forestry plantations. The authors conclude that both land sharing and land sparing can promote biodiversity without compromising high yields, and that a combination of management strategies at different spatial scales, including the maintenance of forest connectivity, may most effectively safeguard both biodiversity and livelihood security.

In Chapter 8, Peh and colleagues examine the impact of forest fragmentation on ecological functioning. They found that the severity and direction of fragmentation effects are influenced by the individual species involved, the environment and other contributory factors (e.g. surrounding matrix habitat quality), and while processes heavily influenced by larger species are likely to respond negatively to fragmentation, others influenced by a wider range of biotic and abiotic factors are less predictable. In addition, cascading effects through the trophic levels may result in the impacts on one trophic level affecting other parts of the ecosystem, and the impacts of fragmentation are often exacerbated by synergistic effects such as interactions with the matrix and increased hunting pressures within fragments.

Forest habitats provide a multitude of economic and ecological services from local to global scales. With over 2 billion rural poor depending on forests to some degree for their livelihoods, understanding how fragmentation might disrupt the provisioning of these livelihoods is a global challenge. In Chapter 9, Garcia and Feintrenie tackle this knotty problem. Their analysis of the impacts of forest fragmentation on livelihoods starts with a careful look at definitions of fragmentation and drivers from both a social and environmental perspective. Using the forest transition curve as a heuristic frame, this chapter presents the myriad of ways that forest fragmentation impacts on livelihoods, from increased wildlife–human conflict to changes in fire frequencies as forest edge increases. Fragmentation can have positive benefits for forest-based livelihoods, and balancing these trade-offs in future forest landscapes is central to multiple-use landscape management and ensuring rapid development in countries while maintaining their forest cover.

Around 75% of forests globally are controlled by governments. In Chapter 10, Barr and colleagues examine the critical role that governance plays in shaping tropical forest fragmentation. Identifying the policy instruments and political challenges to reverse forest fragmentation is perhaps nowhere more urgent than in the tropics, where industrial logging and agro-industrial crops increasingly compete for forested land. This chapter presents the historical context of 'political forests' and explores the challenges of establishing novel sustainable regulations for logging concessions, due to corruption and limited political will. Forest fragmentation is not only driven by but has driven changes in policies, institutional structure and markets, and innovative and effective governance mechanisms are required to both slow and reverse fragmentation. The importance of extra-sectoral governance processes that incentivize forest fragmentation are increasingly globalized. Establishing forest governance that has meaningful forest tenure reform and innovative landscape-scale forest management policies that consider multiple objectives are urgently required.

Finally, in Chapter 11, we provide a summary of the major challenges to reversing forest fragmentation and a synthesis of the major topics covered in this book. We end with an optimistic outlook of the wide range of approaches that are being used to reverse forest fragmentation. Although this provides hope for the future of global forest landscapes, we still require a major shift in thinking from government bodies and industry. We hope that the pages of this book will go some way to ensuring the long-term persistence of forest landscapes that sustain a rich and fruitful habitat for generations to come.

2 The History of Deforestation and Forest Fragmentation: A Global Perspective

Shonil Bhagwat*
The Open University, Milton Keynes, UK; University of Oxford, Oxford, UK; Linacre College, Oxford, UK

Abstract
This chapter draws on long-term ecological studies from across the globe to reconstruct the history of deforestation and forest fragmentation. It includes studies from a variety of forest biomes focusing on the last 11,500 years (the Holocene), paying particular attention to forest fragmentation as a result of expanding agricultural frontiers. Piecing together evidence from a number of case studies from around the world, this chapter attempts to provide a timeline of the earliest occurrence of forest fragmentation and its extent. While there are a large number of studies from Europe and Asia and a few from the Americas, Australasia and Eurasia, limited information is available from tropical Americas, Africa, and South and South-east Asia. This chapter indicates the need for further research of the literature and a global meta-analysis to identify the evidence for historical deforestation and forest fragmentation. The chapter also highlights the need for long-term ecological research and palaeoecological investigations to enhance our understanding of the historical timeline of deforestation and forest fragmentation in the tropics.

Historical Patterns of Deforestation and Forest Fragmentation

Humans are known to have deforested landscapes in many regions of the world throughout the Holocene – a time period spanning the last 11,500 years – causing a decline in forest cover, consequent loss of biodiversity and extinction of iconic species (Turvey, 2009). The history of deforestation, however, varies from region to region. For example, while it is widely accepted that anthropogenic land clearance for agriculture and grazing led to deforestation in many European landscapes during the mid-Holocene (Carcaillet *et al.*, 2002; Tinner *et al.*, 2005), there is limited understanding of deforestation in other regions of the world. Furthermore, the effects of mid-Holocene climate warming and anthropogenic deforestation are often confounded, making it difficult to separate the impact of human activity and climatic changes (e.g. Berglund, 2003; Martínez-Cortizas *et al.*, 2009). A number of studies based on historical archives of fossilized pollen grains, plant parts and charcoal have, however, shed some light on the anthropogenic modification of landscapes throughout

*shonil.bhagwat@open.ac.uk

the history of agricultural expansion over the last 11,500 years. These studies suggest that anthropogenic deforestation has been one of the most pervasive processes in landscape development in Europe during the Holocene (Kaplan *et al.*, 2009; see Chapter 3, this volume). Many of these studies also report the use of fire as a major driver of deforestation and indicate transformation of vegetation from forest to fire-adapted shrubland or grassland along with widespread evidence of burning, grazing and cultivation (e.g. Watkins *et al.*, 2007; Zhao *et al.*, 2010; Morales-Molino *et al.*, 2011). Similar drivers of deforestation have been reported from other regions of the world, namely Central and South America, Australasia and Eurasia, although the timing of deforestation varies widely within and across regions (e.g. López *et al.*, 2003; Horrocks *et al.*, 2007; Carrillo-Bastos *et al.*, 2012). This chapter attempts to examine these global patterns of deforestation and forest fragmentation. Although natural drivers such as sea-level rise may also lead to forest fragmentation in some parts of the world (e.g. Seychelles, Madagascar), this chapter focuses on deforestation and forest fragmentation driven by humans.

Forest fragmentation is the process of subdividing continuous forest into smaller patches, resulting in the loss of habitat, a reduction in patch size and increasing isolation of patches. The palaeoecological literature is based on interpretation of fossilized archives to determine the extent, timing and causes of 'deforestation' (the loss of forest cover), which goes in tandem with 'forest fragmentation'. As such, this chapter assumes that anthropogenic deforestation takes place through the process of forest fragmentation. The available palaeoecological literature provides an insight into the history of forest loss during the Holocene. To examine the relevant literature, a search was carried out of the ISI Web of Knowledge[SM] for three key words – deforestation, Holocene and human activity. A total of 33 studies published between 1995 and 2012 directly reported deforestation at the study sites (Fig. 2.1) and estimated the timing of this deforestation (Figs 2.2 and 2.3). From these estimates, the earliest reported date of deforestation was recorded from 33 locations around the world.

Although this is not an exhaustive review of the literature, the literature surveyed points to some general global patterns and indicates data gaps in enhancing our understanding of global deforestation and forest fragmentation. A large majority of studies (22 in total) were carried out in Europe – five

Fig. 2.1. Location of 33 site-specific studies from the palaeoecological literature reporting the earliest periods of deforestation and forest fragmentation. The majority of the studies (22 in total) were from Europe, seven were from Asia and the remaining four were from the Americas, Australasia and Eurasia.

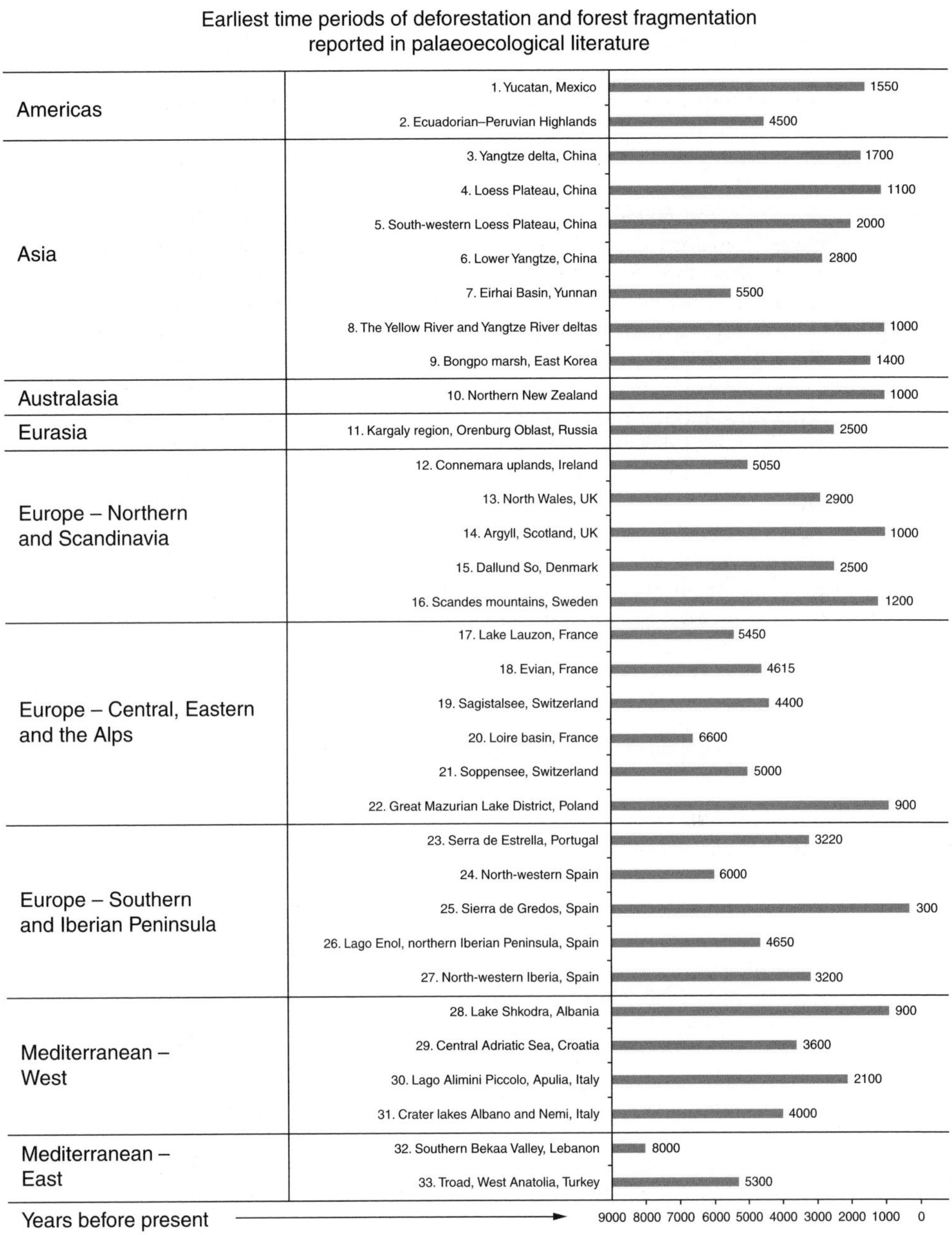

Fig. 2.2. Earliest reported time period of deforestation in 33 case studies from the palaeoecological literature. Most studies reported the expansion of agricultural frontiers as the main cause of deforestation. The dates range from 300 to 8000 years before present. The sources of these studies are as follows: 1, Carrillo-Bastos *et al.* (2012); 2, Weng *et al.* (2004); 3, Wang *et al.* (2011); 4, Zhang *et al.* (2010); 5, Zhao *et al.* (2010); 6, Atahan *et al.* (2008); 7, Yang *et al.* (2005); 8, Saito *et al.* (2001); 9, Park *et al.* (2012); 10, Horrocks *et al.* (2007); 11, Lopez *et al.* (2003); 12, Huang (2002); 13, Watkins *et al.* (2007); 14, Macklin *et al.* (2000); 15, Bradshaw *et al.* (2005); 16, Karlsson *et al.* (2009); 17, Argant *et al.* (2006); 18, Guiter *et al.* (2005); 19, Wick *et al.* (2003); 20, Cyprien *et al.* (2004); 21, Lotter (2001); 22, Wacnik *et al.* (2012); 23, van der Knaap and van Leeuven (1995); 24, Kaal *et al.* (2011); 25, Lopez-Merino *et al.* (2009); 26, Moreno *et al.* (2011); 27, Morales-Molino *et al.* (2011); 28, Zanchetta *et al.* (2011); 29, Oldfield *et al.* (2003); 30, Di Rita and Magri (2009); 31, Guilizzoni *et al.* (2002); 32, Hajar *et al.* (2010); 33, Riehl and Marinova (2008).

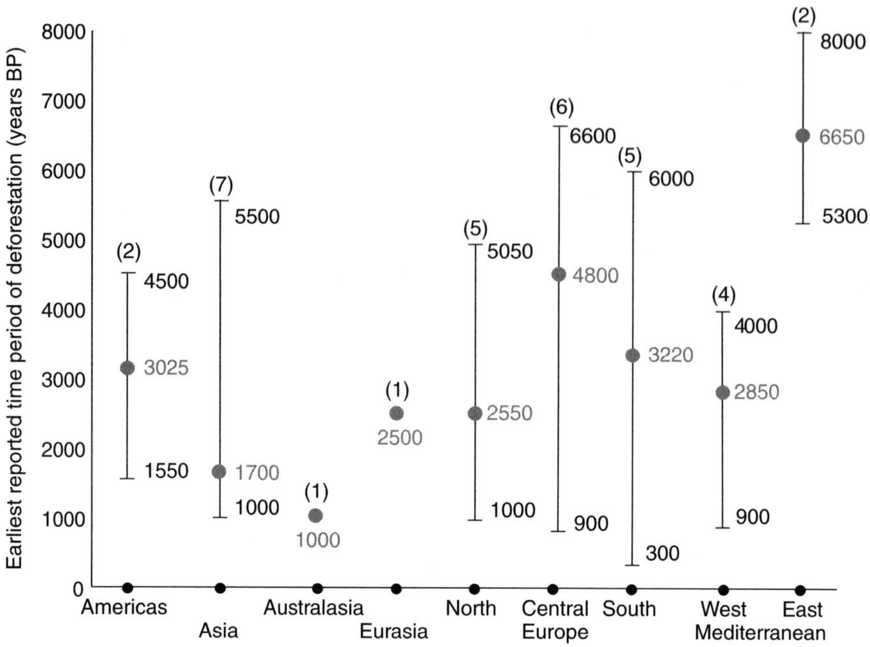

Fig. 2.3. Summary of the earliest reported timeline of deforestation in 33 site-specific palaeoecological studies across the globe. The end lines represent the range of years reported, the grey points on each line represent median YBP (years before present), and numbers in parentheses refer to the number of studies from that region reported in this chapter (see Fig. 2.2 for earliest reported time periods of deforestation in each study).

in northern Europe and Scandinavia, six in Central or Eastern Europe and the Alps, and five in southern Europe including the Iberian Peninsula. Four studies were carried out in the western and two in the eastern Mediterranean region. Seven studies were carried out in Asia – six in China and one in South Korea. There were two studies from the Americas: one in Central America and one in the south. One study was found from Australasia and one from Eurasia. There are a number of other studies that investigated deforestation or forest fragmentation using other methodological approaches (e.g. genetic analysis of single species or groups of species) or other, more recent, timescales. The focus of this chapter, however, is on landscape-scale deforestation caused by human presence and therefore the chapter focuses on the 33 examples identified in the literature.

These studies provide site-specific information on deforestation at 33 locations around the world and help us to gain an insight into the rough timeline of deforestation. This timeline is useful to understand how long the process of deforestation and forest fragmentation has operated in regions around the world. These studies also indicate that the historical timeline of deforestation might be very different at sites in one region, and therefore deforestation and fragmentation are site-specific phenomena. While historical deforestation in Europe has been widely reported, there are few studies from Africa and South-east Asia, and only two from the Americas, meaning that our understanding of the historical timeline of tropical deforestation will remain inadequate until long-term ecological studies have been carried out in these regions. With these caveats in mind, this chapter describes the historical timeline of deforestation in Europe and Asia and parts of the Americas, Australia and Eurasia based on a synthesis of 33 case studies from the palaeoecological literature.

Regional Patterns

The studies reviewed indicate that the timeline of deforestation varies from region to region. What these studies also indicate is that the scale and extent of deforestation are very different. Many of these studies provide an insight into land-use practices by prehistoric cultures and enable a greater understanding of the process of deforestation. For example, some of the studies have provided evidence of large-scale landscape transformation as a result of the expansion of agricultural frontiers, while others indicate small-scale deforestation creating smaller openings in the forest cover. These studies therefore help to enhance our understanding of land-use practices and their effect on forest cover at various sites around the world.

Africa

There is a small body of palaeoecological literature from Africa (e.g. Brncic *et al.*, 2007; Virah-Sawmy *et al.*, 2010), although it does not focus explicitly on identifying the earliest onset of deforestation and forest fragmentation during the Holocene. Quéméré *et al.* (2012) used genetic data to show that the origin of open habitats in Madagascar was pre-human, questioning the prevalent narrative of anthropogenic deforestation of the region. It is likely, however, that anthropogenic deforestation and forest fragmentation on a much smaller scale took place across Africa throughout the history of human presence on the continent.

Americas

Tchir *et al.* (2012) argued that deforestation in the USA and Canada was the result of forest removal for agriculture starting after the arrival of the first European settlers. In the 19th and early 20th centuries, more specialized forms of agriculture were practised, and by the early 20th century most of the deforestation had already taken place. Palaeoecological studies from the pre-European period in North America are rare, but available evidence suggests that landscape modifications in North America started well before Europeans arrived on the continent (Stinchcomb *et al.*, 2011). The study indicates widespread landscape-scale changes during the Little Ice Age, 700–1000 years before present (YBP). Similarly, a large body of literature on the historical landscape development on the Yucatan Peninsula, Mexico, suggests that the Mayan cultures who inhabited this region started large-scale deforestation during the late Holocene (Abrams and Rue 1998; Dunning *et al.*, 2002; Leyden, 2002; Turner *et al.*, 2003). Carrillo-Bastos *et al.* (2012) carried out a spatial analysis of pollen records from the Yucatan Peninsula and, based on their analysis, they provide an insight into the resource use of the Mayan culture. They suggested that deforestation at this site on the Yucatan Peninsula started around 1550 YBP, but this was often on a much smaller spatial scale than assumed by other authors. They speculated that most of the Yucatan Peninsula remained forested, with only small openings in the forest cover until about AD 450, just before the Spanish conquest, when large-scale deforestation ensued.

Weng *et al.* (2004) suggested, however, that deforestation in the highlands of Ecuador and Peru predated deforestation in the Yucatan Peninsula by several centuries. Based on pollen analysis, they demonstrated that pollen of various tree taxa (*Alnus*, *Podocarpus* and *Hedyosmum*) decreased at the same time as an increase in pollen of anthropogenic indicators, such as pollen grains from *Ambrosia* and those from plants in the families Chenopodiaceae and Amaranthaceae. A similar trend is seen in other records, namely from Marcacocha and Baja (Hansen and Rodbell, 1995; Chepstow-Lusty *et al.*, 1996, 2003), suggesting that this deforestation phenomenon around 4500 YBP might have been more widespread than the highlands of Ecuador and Peru. The cause of this deforestation might have been the onset of agriculture in this region coupled with dry and warm climatic conditions during the mid-Holocene. *Alnus* populations are known to have recovered 2000–1000 YBP, suggesting that land management for agriculture along with climatic warming around 4500 YBP might have exacerbated the tree decline, but that the forest cover subsequently increased across the region.

Asia

A number of palaeoecological studies reported in the literature over the last decade or so suggest that the human impact on forest vegetation in Asia, particularly in China, might have started as early as 5000 YBP, and that human activities might have intensified deforestation in China (particularly on the Loess Plateau, in the Yellow River catchment and in the Erhai Lake drainage basin in Yunnan province) over the last 2000 years (Saito *et al.*, 2001; Huang *et al.*, 2002; Yi *et al.*, 2003a,b; Makohonienko *et al.*, 2004; Li *et al.*, 2006). The fossil pollen data of Ren (2000) from north China and the upper-middle Yangtze basin indicated no obvious forest decline in the Yangtze basin during the mid- to late Holocene, while severe decline of arboreal pollen is known to have occurred elsewhere since 5000 YBP. Yasuda *et al.* (2004) reported large-scale deforestation of evergreen oak forests between 5300 and 4500 YBP in the middle Yangtze basin.

Zhang *et al.* (2010) reconstructed vegetation dynamics and human activities from lake sediment archives in the western Loess Plateau, China. Based on the first appearance of cereal-type pollen grains (indicating the beginning of settled agriculture), they suggested that cereal cultivation might have expanded in the upland regions of the Loess Plateau and that this was accompanied by large-scale deforestation around 1100 YBP. Since then, several distinct phases of deforestation are recorded in the sediment archives, and they correspond to the documented history of human activities on the Loess Plateau. The first significant deforestation between 1100 and 750 YBP corresponds with the late Tang and Song Dynasties, a period when agriculture is known to have flourished and the human population is known to have increased and concentrated in fertile regions (Xu, 1998). During the Yuan and the early Ming Dynasties, the population is known to have declined, leading to abandonment of agricultural land, resulting in expansion of forests between 700 and 50 YBP (Saito *et al.*, 2001). The final substantial forest clearance is known to have occurred around 50 calibrated YBP. Vegetation history inferred from pollen records generally agrees with the deforestation records based on the historical documents in the Loess Plateau. During the late Ming Dynasty and Qing Dynasty (from 350 YBP onwards), another large-scale clearance of forest is known to have occurred due to intensification of agriculture across the Loess Plateau.

The study of Zhao *et al.* (2010) from the Liupan Mountains in the south-western Loess Plateau in central China traced human activities over 6200 years. Their results from pollen and charcoal analysis in conjunction with archaeological and historical evidence suggested that human activities over the last 2000 years have significantly accelerated deforestation, which might have started during a drying climate in the mid-Holocene. Their evidence of *Artemisia* and Poaceae pollen grains along with abundant charcoal in sediment layers suggest that people used the slash-and-burn practice of forest clearance and made use of fire for land clearance. Interpreted in conjunction with historical archives, their results suggest that deforestation in the Liupan Mountain region was very limited in extent before the Qin Dynasty (*c.*2150 YBP), but increased after the Han Dynasty (*c.*2170 YBP) due to widespread and intensive agriculture practices. After the Tang and Song Dynasties (*c.*700 YBP), deforestation occurred on a much larger scale in this region (Wen and Wen, 2006). These results also suggest that, although the slash-and-burn technique was commonly used to clear vegetation and to practise agriculture around 2000 YBP, fire was less commonly used between 1000 and 500 YBP because farming expanded on land that was already cleared.

Yang *et al.* (2005) provided pollen evidence of early human activities in the Eirhai basin, Yunnan province, China. They suggested that deforestation started with selective clearance of forests in certain parts of the basin around 5500 YBP, and speculated that, around this time period, montane forests were lost, while secondary pine woodland expanded across the basin. This change was also accompanied by an increase in herbs in relation to trees, indicating primitive agriculture grazing in the study region. This study suggests that the time period of anthropological impact on the Eirhai basin predates the oldest archaeological records by about 1500 years and therefore provides an insight into land management practices of prehistoric people in this region.

Atahan *et al.* (2008) reviewed the existing sedimentary records of environmental changes and early agriculture in the lower Yangtze, China. They provided possible evidence of human activity leading to deforestation (possibly related to agriculture) as early as 7000 YBP; clearer evidence of human activity went back to 4700 YBP, while widespread human impacts appeared between 2800 and 2200 YBP. In the lower Yangtze River basin, this latter period is known to have been associated with technological advances in agriculture and increased urbanization. Wang *et al.* (2011), on the other hand, indicated that, in the Yangtze River basin, the impacts of human activity may be more recent and that large-scale deforestation might have started only around 1700 YBP. Their findings were based on a number of sediment archives, some of which went back to 6250 YBP. Sediment accumulation rates and the magnetic properties of these archives suggested that major deforestation might have started in the Yantze River basin around 1700 YBP and that it intensified after about 1800 YBP. They speculated that migration of human populations from northern China into the upper stretches of the Yantze River basin around 1700 YBP might have led to deforestation in the uplands, resulting in increased sediment discharge into the Yangtze River. The study of Saito *et al.* (2001) in the Yellow River and the Yangtze River deltas reported an abrupt increase in sediment discharge into both rivers around 1000 YBP. They attributed this to a sharp increase in deforestation and cultivation on the Loess Plateau, suggesting linkages between intensifying human activity in the uplands and its effects downstream.

According to Park *et al.* (2012), deforestation on the east coast of South Korea is much more recent in comparison with the timescale of forest loss in China but follows a similar cause-and-effect chronology between deforestation in the uplands and its effects in terms of sediment discharge in the lowlands. Based on multiproxy evidence for late Holocene anthropogenic changes at Bongpo marsh, they showed that the marsh began to develop around 2660 YBP, and agricultural disturbance around the study site remained limited until 2610. They reported that the area of deforestation increased many fold between 2610 and 140 YBP, possibly because the land use changed from lowland rice agriculture to upland cultivation, leading to increased deforestation in the uplands.

There are fewer records of early deforestation and forest fragmentation from South Asia, but Bhagwat *et al.* (2012), based on a sediment archive from South India, indicated that the earliest evidence of landscape burning was found around 3500 YBP. However, this study also indicated that other factors, such as a decrease in the strength of the monsoon, might have had an effect on landscape development in this part of South Asia. Other parts of South Asia may have had a variable history of deforestation and forest fragmentation, and therefore the record for the earliest evidence of deforestation may not be consistent across the region.

Australasia

The study by Horrocks *et al.* (2007) showed that the beginning of large-scale deforestation in northern New Zealand dates back to approximately 1000 YBP or later. Until this time, conifer–hardwood forests dominated by podocarps (especially *Dacrydium*) prevailed in the region. This finding concurs with previous pollen studies (e.g. Elliot *et al.*, 1997; Horrocks *et al.*, 2001; Newnham *et al.*, 2004) suggesting that regional Polynesian deforestation occurred at a similar time to that of much of New Zealand, generally 800–600 YBP (McGlone, 1983; McGlone and Wilmshurst, 1999). Therefore, 800–600 YBP appears to be the time period of widespread changes in the landscape, although early deforestation might have started around 1000 YBP.

Eurasia

Lopez *et al.* (2003) carried out a study based on pollen analysis of Novienki peat bog in the Kargaly region of Russia, which indicated that deforestation of birch forests in this region started around 2500 YBP, a period corresponding to the Bronze Age in this region. They attributed this deforestation to metallurgical activities rather than to agriculture.

Europe

Northern Europe and Scandinavia

In northern Europe, various studies have reported the timescale of deforestation in Ireland, Wales and Scotland. The study by Huang (2002) in the Connemara uplands in western Ireland indicated that deforestation in this region started between 5050 and 4000 YBP. Watkins *et al.* (2007) studied the vegetation history of the Arfon Platform in North Wales, UK, and found that increased use of fire and the spread of grassland began in this region around 2600 YBP, although the first cereal grains were recorded around 2900 YBP. This suggests that agriculture spread rapidly in this region between 2900 and 2600 YBP. Macklin *et al.* (2000) interpret human–environment interactions during the Holocene in the Oban area of Argyll, Scotland. Their analysis showed that, up until approximately 1000 YBP, agricultural communities had comparatively little impact on the environment, although cultivation of cereal grains might have started much earlier – around 5000 YBP. After 1000 YBP, rapid and permanent deforestation took place, possibly linked to the development of a distinctive land-use strategy and settlement pattern that survived until the Industrial Revolution in approximately 200 YBP. According to Macklin *et al.* (2000) the underlying causes of deforestation after 1000 YBP might include a population increase and a growing requirement for agricultural land and timber for fuel and building, soil decline associated with climate change alongside human mismanagement of soil, extraction of timber for export to other regions or for boat and harbour construction during the period of Norse influence, and deliberate removal of vegetation cover as an instrument of warfare.

In Scandinavia, two studies reported the earliest onset of deforestation – one from Dallund So in Denmark (Bradshaw *et al.*, 2005) and another from the Swedish Scandes Mountains (Karlsson *et al.*, 2009). Bradshaw *et al.* (2005) indicated that major deforestation started in Denmark around 2500 YBP, until when there was very little evidence of deforestation in this region. According to Karlsson *et al.* (2009), deforestation in the Swedish Scandes Mountains began only in the 8th century, approximately 1200 YBP. Until then, the region around their study site was forested with *Betula* trees. This study compared its findings with another study from the same region and noted differences in the magnitude of vegetation change and post-disturbance vegetation development, and found that site-specific factors played a disproportionately important role in determining the patterns of deforestation and vegetation recovery.

Central and Eastern Europe and the Alps

Argant *et al.* (2006) reported that, at Lake Lauzon in France, the first impact of human activity on forest cover occurred around 7400 YBP, possibly with isolated clearings and evidence of the occurrence of cereal pollen grains. Clearer evidence of agriculture – and therefore large-scale deforestation – however, emerged only around 5450 YBP, suggesting that arable agriculture and grazing became much more widespread at this time. In the Evian area of France, deforestation might have ensued slightly later, around 4615 YBP (Guiter *et al.*, 2005), but the causes of this deforestation are similar, namely arable agriculture and grazing. Tombs and vestiges of villages around this lake indicate that the region was intensively and recurrently occupied by people during the suggested time period of deforestation, agriculture and grazing. In the Loire River basin in western France, deforestation might have started much earlier, around 6600 YBP (Cyprien *et al.*, 2004). The causes of this deforestation might have been animal grazing, because evidence suggests that arable agriculture did not start until around 5600 YBP, when wheat, rye, buckwheat and flax were cultivated. The patterns of deforestation also suggest that chestnut and walnut might have been exploited in the Loire Valley region at about 4600 YBP, while vineyards started around 2000 YBP onwards – also a time period when deforestation reached its peak. The reports of first deforestation from central Switzerland also date back to around 5000 YBP, when the evidence of first cereal grains is found (Lotter, 2001). A more large-scale loss of forest cover, however, did not occur until the Middle Ages, approximately 1500 YBP, when the forest ecosystem changed drastically, giving way to much more open landscapes. Evidence suggests that after World War II, agriculture intensified even further, as the landscape retained its open character until today.

In the Swiss Alps, extensive deforestation might have started around 3700 YBP due to intensive pasturing and grazing (Wick *et al.*, 2003), although the human impact began 4400 YBP. Whereas the impact on forest around 4400 YBP was moderate, deforestation and grazing pressure since 3700 led to lowering of the forest limit in the Swiss Alps at least 200–300 m below its natural level (Wick and Tinner, 1997). Later, during the Middle Ages (*c.*1500 YBP), the pollen record suggests a sharp decline in tree pollen grains such as *Picea*, *Abies* and *Alnus viridis* and sharp increase in pollen grains of *Plantago alpina*, Compositae and *Potentilla*-type plants, indicating a sharp transition from a tree-covered to a treeless landscape supporting large-scale cattle breeding, pasturing and grazing (Wick *et al.*, 2003).

In the Great Mazurian Lake District of Poland, evidence of deforestation is much more recent in comparison with central Europe or the Alps. Wacnik *et al.* (2012) dated large-scale deforestation to approximately 900 YBP as a result of agricultural activity. Similar findings are reported from previous studies in other lake basins in the same region (Bukowski *et al.*, 1965 and Filbrandt-Czaja, 2000, cited by Wacnik *et al.*, 2012) and these all indicate the beginnings of deforestation due to permanent human settlement linked with agriculture, grazing and other subsistence activities.

Southern Europe and Iberian Peninsula

Evidence of small-scale local deforestation and grazing in Portugal dates back to 4550 YBP, but large-scale deforestation is known to have started only after 3220 YBP (van der Knaap and van Leeuwen, 1995). From about 955 YBP, the evidence of anthropogenic pressure is suggested to have increased to such an extent that forest was seen to disappear and the topsoil eroded. According to van der Knaap and van Leeuwen (1995), the causes of this deforestation were grazing, burning, agriculture and large-scale plantation of pines. In the uplands of the Teleno Mountains in the north-western Iberian Peninsula, Morales-Molino *et al.* (2011) also dated the earliest deforestation to around 3220 YBP. They indicated that, between 4500 and 3200 YBP, *Pinus* and *Betula* trees dominated the landscape, but around 3200 YBP, pine–birch forests were replaced with heathlands and grasslands, which persisted all the way up to the Roman period until approximately 1500 YBP. Morales-Molino *et al.* (2011) suggest that this abrupt deforestation was caused by fire, grazing, and/or mining activities linked to the exploitation of the metal resources of the Teleno Mountains. Subsequently, *Betula* forests re-established around 1250 YBP, possibly due to a temporary decrease in human activities. Between 300 and 200 YBP, the regional population density is known to have increased. This increase might have caused an increase in fire activity, which once again replaced the birch forests with heathlands.

In the northern Iberian Peninsula, one study dates the onset of deforestation to 4650 YBP (Moreno *et al.*, 2011). They suggested that intense human activity for animal grazing might have led to deforestation and formation of the resulting landscapes of alpine grassland and meadows.

The evidence of deforestation and landscape burning in the wider north-western Spain goes much further back (*c.*6000 YBP), as suggested by the presence of pollen grains of *Plantago lanceolata* and *Urtica dioica* in some of the sediment archives from north-western Spain, indicating the presence of pastureland as early as 6000 YBP (Kaal *et al.*, 2011).

In Sierra de Gredos in central Spain, on the other hand, deforestation appears to be much more recent and dates back only to the 18th century, approximately 300 YBP, according to López-Merino *et al.* (2009). However, it is commonly known that the whole of the central Iberian Peninsula suffered from sudden, rapid and irreversible deforestation of mountain pine forests (Ruiz-Zapata *et al.*, 1996, 2006, 2007, cited by López-Merino *et al.*, 2009). Around 300 YBP, however, this became much more intensified with continued human activity in the mountain pine forests, the use of fire and intensification of grazing practices leading to progressive deforestation, and the expansion of fire-prone scrub. During the 19th century, grazing is known to have intensified further, causing more widespread deforestation (Rubiales *et al.*, 2007).

Mediterranean

Western Mediterranean

Zanchetta *et al.* (2011) reported that, in the area around Lake Shkodra on the border between Albania and Montenegro, deforestation might have started around 900 YBP. This is based on the evidence of a significant decrease in the percentage of arboreal pollen and in pollen concentration from a sediment archive from this lake. In the wider western Mediterranean area, however, human impact is known to have started much earlier, as suggested by a high-resolution sediment core from the central Adriatic Sea (Oldfield *et al.*, 2003). In this sediment archive, the most obvious episodes of deforestation are noted starting 3600 YBP, as evidenced by an acceleration of mass sedimentation in the Adriatic Sea reflecting widespread forest clearance and cultivation. A similar episode of deforestation was noted again around 700 YBP, possibly linked to forest clearance during the medieval period.

In Lago Alimini Piccolo lake, Apulia, south-east Italy, Di Rita and Magri (2009) reported that, between 5200 and 4350 YBP, a dense evergreen oak forest dominated the landscape. Small openings in the forest were recorded between 4350 and 3900 YBP followed by forest expansion between 3900 and 2100 YBP, characterized by an increase in olive trees, suggesting cultivation of olive groves without widespread deforestation in the region. During the Roman occupation period (2100–1500 calibrated YBP), however, a significant opening of forest was observed. After 1500 YBP, human activities are reported to have caused a further decrease in the natural woodland and expansion of olive groves. At many other Italian sites that are at a similar latitude to Lago Alimini Piccolo lake, however, widespread opening and clearing of vegetation was noted during a 500-year period between 3950 and 4450 YBP, suggesting that deforestation in other regions might be more large-scale.

In central Italy, Guilizzoni *et al.* (2002) reported Holocene reconstruction of the lake environment at two crater lakes, Albano and Nemi. They indicated that two events during the course of history of the lakes – the first around 3500–3000 YBP during the Bronze Age and the second around 1800 YBP during the Roman period – when deforestation might have occurred. The fact that these changes were synchronous in both lakes suggests that deforestation operated across the landscape in central Italy and that it took place at a similar time period over the entire region.

Eastern Mediterranean

As a cradle of agriculture, the history of deforestation in the eastern Mediterranean naturally goes back a long time. In the Fertile Crescent, for example, pollen records from Israel (Baruch, 1990; Schwab *et al.*, 2004) indicate cultivation of olive and simultaneous removal of oak-dominated forests from around 6000 YBP. In Troad in Western Anatolia, Turkey, Riehl and Marinova (2008) reported a widespread decline in deciduous oak forests around 5300 YBP (Bottema *et al.*, 2001, cited by Riehl and Marinova, 2008). Olive cultivation was recorded in Syria (Niklewski and van Zeist, 1970; Yasuda *et al.*, 2000) and, although the dating of this cultivation is uncertain (Meadows, 2005), it is likely to be several thousand years old. In Mount Lebanon, evidence of deforestation from 5500 YBP exists, and this evidence is related to the removal of cedar forests by the Egyptians (Loffet, 2004, cited by Hajar *et al.*, 2010). Hajar *et al.* (2010) reported vegetation and environmental changes in the

Southern Bekaa Valley, which indicate major deforestation events as far back as 8000 YBP, possibly related to the beginning of agriculture. Archaeological records from the same region also indicate a small-scale human impact on forests during the Neolithic period (11500–8500 YBP), as evidenced by flint tools such as axes and chisels. This small-scale human impact gradually increased, and by about 8000 YBP the vegetation in the Bekaa Valley underwent transformation as elsewhere in Eurasia around the same time (Ruddiman, 2003), although the scale of deforestation in the eastern Mediterranean was much greater, possibly due to higher population densities in comparison with elsewhere in Eurasia at that time.

Conclusion

Although these 33 studies and other existing literature do not provide a sufficiently widespread data set to make generalizations about global deforestation and forest fragmentation, a number of insights can be drawn:

1. A large majority of studies of historical deforestation come from Europe, where deforestation and fragmentation appears to have been one of the most pervasive processes in landscape development during the last 11,500 years. To enable large-scale land clearing for agriculture and grazing, humans have used fire as a common tool. As a result, landscapes have been transformed from forest to fire-adapted shrubland or grassland.

2. Drivers of deforestation on other continents, namely Central and South America, Australasia and Eurasia, where the remaining studies come from, are similar to those in Europe. Although the timing of deforestation has varied widely within and across regions, the process of deforestation appears to have been similar across regions, with most deforestation taking place initially for the development of agriculture and subsequent intensification of land use.

3. The eastern Mediterranean region has some of the earliest reported dates of deforestation due to its status as a cradle of agriculture. In general, large-scale deforestation in the 'old world' predates deforestation in other parts of the world, namely the Americas and Australasia, where large-scale deforestation ensued only after the arrival of the Europeans. Asia, particularly China, also has a long history of agriculture, but most studies from this region report small-scale deforestation, with large-scale land clearing starting only during the last two millennia in most sites. Studies from the African and South Asia tropics also indicate small-scale deforestation for much of the historical time period.

Based on the relative 'antiquity' of deforestation across continents, future work should identify regions that have remained resilient in the face of anthropogenic changes and regions that have not.

While this chapter has attempted to unravel some of the global patterns of deforestation, it has also highlighted the need for further research into global deforestation over millennial timescales. The chapter has highlighted comparative patterns of deforestation around the world based on a small subset of the literature, but there is scope for a much broader literature survey, review and meta-analysis in order to generate a more comprehensive picture. In particular, studies from Africa, South Asia and South-east Asia are needed; this is also where the oldest evidence of deforestation might be found due to the long history of human habitation of these regions.

References

Abrams, E.M. and Rue, D.J. (1998) The causes and consequences of deforestation among the prehistoric Maya. *Human Ecology* 16, 377–395.

Argant, J., López-Sáez, J.A. and Bintz, P. (2006) Exploring the ancient occupation of a high altitude site (Lake Lauzon, France): comparison between pollen and non-pollen palynomorphs. *Review of Palaeobotany and Palynology* 141, 151–163.

Atahan, P., Itzstein-Davey, F., Taylor, D., Dodson, J., Qin, J., Zheng, H. and Brooks, A. (2008) Holocene-aged sedimentary records of environmental changes and early agriculture in the lower Yangtze, China. *Quaternary Science Reviews* 27, 556–570.

Baruch, U. (1990) Palynological evidence of human impact on the vegetation as recorded in Late Holocene lake sediments in Israel. In: Bottema, S., Entjes-Nieborg, G. and van Zeist, W. (eds) *Handbook of Man's Role in the Shaping of the Eastern Mediterranean Landscape*. Balkema, Rotterdam, pp. 283–293.

Berglund, B.E. (2003) Human impact and climate changes – synchronous events and a causal link? *Quaternary International* 105, 7–12.

Bhagwat, S., Nogué, S. and Willis, K.J. (2012) Resilience of an ancient tropical forest landscape to 7500 years of environmental change. *Biological Conservation* 153, 108–117.

Bottema, S., Woldring, H. and Kayan, I. (2001) The late quaternary vegetation of Western Turkey. In: Roodenberg, J. and Thissen, L. (eds) *The Ilipinar Excavations II*. Nederland Intituut voor Nabije Oosten, Leiden, pp. 327–356.

Bradshaw, E.G., Rasmussen, P. and Odgaard, B.V. (2005) Mid- to late-Holocene land-use change and lake development at Dallund So, Denmark: synthesis of multiproxy data, linking land and lake. *Holocene* 15, 1152–1162.

Brncic, T.M., Willis, K.J., Harris, D.J. and Washington, R. (2007) Culture or climate? The relative influences of past processes on the composition of the lowland Congo rainforest. *Philosophical Transactions of The Royal Society B: Biological Sciences* 362, 229–242.

Bukowski, Z., Dabrowski, J., Dabrowski, M. and Odoj, R. (1965) Wyniki podwodnych badań archeologicznych w jez. Piłakno, pow. Mrągowo, w 1962 roku [Results of underwater archaeological research in the Lake Piłakno, distr. Mrągowo, in 1962]. *Sprawozdania Archeologiczne* 17, 100–113.

Carcaillet, C., Almquist, H., Asnong, H., Bradshaw, R.H.W., Carrión, J.S., Gaillard, M.-J., Gajewski, K., Haas, J.N., Haberle, S.G., Hadorn, P., Müller, S.D., Richard, P.J.H. *et al.* (2002) Holocene biomass burning and global dynamics of the carbon cycle. *Chemosphere* 49, 845–863.

Carrillo-Bastos, A., Islebe, G.A. and Torrescano-Valle, N. (2012) Geospatial analysis of pollen records from the Yucatan peninsula, Mexico. *Vegetation History and Archaeobotany* 21, 429–437.

Chepstow-Lusty, A.J., Bennett, K.D., Switsur, V.R. and Kendall, A. (1996) 4000 years of human impact and vegetation change in the central Peruvian Andes – with events parallelling the Maya record? *Antiquity* 70, 824–833.

Chepstow-Lusty, A., Frogley, M.R., Bauer, B.S., Bush, M.B. and Herreras, A.T. (2003) A late Holocene record of arid events from the Cuzco region, Peru. *Journal of Quaternary Science* 18, 491–502.

Cyprien, A.L., Visset, L. and Carcaud, N. (2004) Evolution of vegetation landscapes during the Holocene in the central and downstream Loire basin (Western France). *Vegetation History and Archaeobotany* 13, 181–196.

Di Rita, F. and Magri, D. (2009) Holocene drought, deforestation and evergreen vegetation development in the Central Mediterranean: a 5500 years record from Lago Alimini Piccolo, Apulia, Southeast Italy. *Holocene* 19, 295–306.

Dunning, N.P., Luzzadder-Beach, S., Beach, T., Jones, J.G., Scarborough, V. and Culbert, T.P. (2002) Arising from the Bajos: the evolution of a Neotropical landscape and the rise of Maya civilization. *Annals of the Association of American Geographers* 92, 267–283.

Elliot, M.B., Striewski, B., Flenley, J.R., Kirkman, J.H. and Sutton, D.G. (1997) A 4300 year palynological and sedimentological record of environmental change and human impact from Wharau Road Swamp, Northland, New Zealand. *Journal of the Royal Society of New Zealand* 27, 401–418.

Filbrandt-Czaja, A. (2000) Vegetation changes in the surroundings of Lake Dgał Wielki in the light of pollen analysis. In: Kola, A. (ed.) *Studies in Lake Dwellings of West Baltic Barrow Culture*. UMK, Toruń, Poland, pp. 89–99.

Guilizzoni, P., Lami, A., Marchetto, A., Jones, V., Manca, M. and Bettinetti, R. (2002) Palaeoproductivity and environmental changes during the Holocene in central Italy as recorded in two crater lakes (Albano and Nemi). *Quaternary International* 88, 57–68.

Guiter, F., Andrieu-Ponel, V., de Beaulieu, J.-L, Ponel, P., Nicoud, G. and Blavoux, B. (2005) Impact anthropique sur la végétation de la région d'Évian depuis le Néolithique moyen: l'enregistrement pollinique de la Beunaz (971 m n.g.f., Haute-Savoie, France). *Comptes Rendus Biologies* 328, 7, 661–673 (in French).

Hajar, L., Haï dar-Boustani, M., Khater, C. and Cheddadi, R. (2010) Environmental changes in Lebanon during the Holocene: man vs. climate impacts. *Journal of Arid Environments* 74, 746–755.

Hansen, B.C.S. and Rodbell, D.T. (1995) A late-glacial/Holocene pollen record from the eastern Andes of Northern Peru. *Quaternary Research* 44, 216–227.

Horrocks, M., Nichol, S.L., Gregory, M.R., Creese, R. and Augustinus, P.C. (2001) A Holocene pollen and sediment record of Whangape Harbour, Far Northern New Zealand. *Journal of the Royal Society of New Zealand* 31, 411–424.

Horrocks, M., Nichol, S.L., Augustinus, P.C. and Barber, I.G. (2007) Late Quaternary environments, vegetation and agriculture in northern New Zealand. *i* 22, 267–279.

Huang, C.C. (2002) Holocene landscape development and human impact in the Connemara uplands, western Ireland. *Journal of Biogeography* 29, 153–165.

Huang, C.C., Pang, J.L., Huang, P., Hou, C. and Han, Y. (2002) High-resolution studies of the oldest cultivated soils in the southern Loess Plateau of China. *Catena* 47, 29–42.

Kaal, J., Carrión Marco, Y., Asouti, E., Martín Seijo, M., Martínez Cortizas, A., Costa Casáis, M. and Criado Boado, F. (2011) Long-term deforestation in NW Spain: linking the Holocene fire history to vegetation change and human activities. *Quaternary Science Reviews* 30, 161–175.

Kaplan, J.O., Krumhardt, K.M. and Zimmermann, N. (2009) The prehistoric and preindustrial deforestation of Europe. *Quaternary Science Reviews* 28, 3016–3034.

Karlsson, H., Shevtsova, A. and Hörnberg, G. (2009) Vegetation development at a mountain settlement site in the Swedish Scandes during the late Holocene: palaeoecological evidence of human-induced deforestation. *Vegetation History and Archaeobotany* 18, 297–314.

Leyden, B.W. (2002) Pollen evidence for climatic variability and cultural disturbance in the Maya lowlands. *Ancient Mesoamerica* 13, 85–101.

Li, Y.Y., Willis, K.J., Zhou, L.P. and Cui, H.T. (2006) The impact of ancient civilization on the northeastern Chinese landscape: palaeoecological evidence from the Western Liaohe River Basin, Inner Mongolia. *The Holocene* 16, 1109–1121.

Loffet, H.C. (2004) Sur quelques espèces d'arbres de la zone Syro-Palestinienne et Libanaise exporté es vers l'Egypte Pharaonique. *Archaeology and History in Lebanon* 19, 10–33 (in French).

López, P., López-Sáez, J.A., Chernykh, E.N. and Tarasov, P. (2003) Late Holocene vegetation history and human activity shown by pollen analysis of Novienki peat bog (Kargaly region, Orenburg Oblast, Russia). *Vegetation History and Archaeobotany* 12, 75–82.

López-Merino, L., López-Sáez, J.A., Alba-Sánchez, F., Pérez-Díaz, S. and Carrión, J.S. (2009) 2000 years of pastoralism and fire shaping high-altitude vegetation of Sierra de Gredos in central Spain. *Review of Palaeobotany and Palynology* 158, 42–51.

Lotter, A.F. (2001) The palaeolimnology of Soppensee (Central Switzerland), as evidenced by diatom, pollen, and fossil-pigment analyses. *Journal of Paleolimnology* 25, 65–79.

Macklin, M.G., Bonsall, C., Davies, F.M. and Robinson, M.R. (2000) Human–environment interactions during the Holocene: new data and interpretations from the Oban area, Argyll, Scotland. *The Holocene* 10, 109–121.

Makohonienko, M., Kitagawa, H., Naruse, T., Nasu, H., Momohara, A., Okuno, M., Fujiki, T., Liu, X., Yasuda, Y. and Yin, H. (2004) Late-Holocene natural and anthropogenic vegetation changes in the Dongbei Pingyuan (Manchurian Plain), northeastern China. *Quaternary International* 123–125, 71–88.

Martínez-Cortizas, A., Costa-Casais, M. and López-Sáez, J.A. (2009) Environmental change in NW Iberia between 7000 and 500 cal BC. *Quaternary International* 200, 77–89.

McGlone, M.S. (1983) Polynesian deforestation of New Zealand: a preliminary synthesis. *Archaeology in Oceania* 18, 11–25.

McGlone, M.S. and Wilmshurst, J.M. (1999) Dating initial Maori environmental impact in New Zealand. *Quaternary International* 59, 5–16.

Meadows, J. (2005) The Younger Dryas episode and the radiocarbon chronologies of the Lake Huleh and Ghab Valley pollen diagrams, Israel and Syria. *The Holocene* 15, 631–636.

Morales-Molino, C., García Antón, M. and Morla, M. (2011) Late Holocene vegetation dynamics on an Atlantic–Mediterranean mountain in NW Iberia. *Palaeogeography, Palaeoclimatology, Palaeoecology* 302, 323–337.

Moreno, A., López-Merino, L., Leira, M., Marco-Barba, J., González-Sampériz, P., Valero-Garcés, B.L., López-Sáez, J.A., Santos, L., Mata, P. and Ito, E. (2011) Revealing the last 13,500 years of environmental history from the multiproxy record of a mountain lake (Lago Enol, northern Iberian Peninsula). *Journal of Paleolimnology* 46, 327–349.

Newnham, R.M., Lowe, D.J., Green, J.D., Turner, G.M., Harper, M.A., McGlone, M.S., Stout, S.L., Horie, S. and Froggatt, P.C. (2004) A discontinuous ca. 80 ka record of Late Quaternary environmental change from Lake Omapere, Northland, New Zealand. *Palaeogeography, Palaeoclimatology, Palaeoecology* 207, 165–198.

Niklewski, J. and van Zeist, W. (1970) A Late Quaternary pollen diagram from northwestern Syria. *Acta Botanica Neerlandica* 9, 737–754.

Oldfield, F., Asioli, A., Accorsi, C.A., Mercuri, A.M., Juggins, S., Langone, L., Rolph, T., Trincardi, F., Wolf, G., Gibbs, Z., Vigliotti, L., Frignani, M. *et al.* (2003) A high resolution late Holocene palaeo environmental record from the central Adriatic Sea. *Quaternary Science Reviews* 22, 319–342.

Park, J., Yu, K.B., Lim, H.S. and Shin, Y.H. (2012) Multi-proxy evidence for late Holocene anthropogenic environmental changes at Bongpo marsh on the east coast of Korea. *Quaternary Research* 78, 209–216.

Quéméré, E., Amelot, X., Pierson, J., Crouau-Roy, B. and Chikhi, L. (2012) Genetic data suggest a natural prehuman origin of open habitats in northern Madagascar and question the deforestation narrative in this region. *Proceedings of the National Academy of Sciences USA* 109, 13028–13033.

Riehl, S. and Marinova, E. (2008) Mid-Holocene vegetation change in the Troad (W Anatolia): man-made or natural? *Vegetation History and Archaeobotany* 17, 297–312.

Rubiales, J.M., García-Amorena, I., Génova, M., Gómez-Manzaneque, F. and Morla, C. (2007) The Holocene history of highland pine forests in a submediterranean mountain: the case of Gredos mountain range (Iberian Central range, Spain). *Quaternary Science Reviews* 26, 1759–1770.

Ruddiman, W.F. (2003) The anthropogenic greenhouse era began thousands of years ago. *Climatic Change* 61, 261–293.

Ruiz-Zapata, M.B., Gil, M.J. and Dorado, M. (1996) Climatic changes in the Spanish Central Zone during the last 3000 BP based on pollinic analysis. In: Angelakis, A.N. and Issar, A.S. (eds*) Diachronic Climatic Impacts on Water Resources with Emphasis on the Mediterranean Region*. Springer-Verlag, Berlin, pp. 9–23.

Ruiz-Zapata, M.B., Gómez, C., López-Sáez, J.A., Gil, M.J., Santiesteban, J.I., Mediavilla, R., Dorado, M. and Valdeolmillos, A. (2006) Detección de la actividad antrópica durante el Holoceno reciente, a través de la asociación de palinomorfos polínicos y no polínicos en dos depósitos higroturbosos (El Berrueco y Rascafría) en la Sierra de Guadarrama, Madrid. *Revista Española de Micropaleontología* 38, 355–366 (in Spanish).

Ruiz-Zapata, M.B., Gómez, C., López-Sáez, J.A., Gil, M.J., Vera, M.S., Mediavilla, R., Domínguez, F. and Santiesteban, J. (2007) Cambios en la vegetación durante el Holoceno reciente en el Valle del Lozoya (Sierra de Guadarrama, Madrid). *Revista Española de Paleontología* 22, 95–102 (in Spanish).

Saito, Y., Yang, Z. and Hori, K. (2001) The Huanghe (Yellow River) and Changjiang (Yangtze River) deltas: a review on their characteristics, evolution and sediment discharge during the Holocene. *Geomorphology* 41, 219–231.

Schwab, M.J., Neumann, F., Litt, T., Negendank, J.F.W. and Stein, M. (2004) Holocene palaeoecology of the Golan Heights (Near East): investigation of lacustrine sediments from Birkat Ram crater lake. *Quaternary Science Reviews* 23, 1723–1731.

Stinchcomb, G.E., Messner, T.C., Driese, S.G., Nordt, L.C. and Stewart, R.M. (2011) Pre-colonial (AD 1100–1600) sedimentation related to prehistoric maize agriculture and climate change in eastern North America. *Geology* 39, 363–366.

Tchir, T.L., Johnson, E. and Nkemdirim, L. (2012) Deforestation in North America: past, present and future. Encyclopedia of Life Support Systems (EOLSS) Regional Sustainable Development Review: Canada and USA – Vol. I. <http://www.eolss.net/Eolss-sampleAllChapter.aspx> (accessed 29 July 2014).

Tinner, W., Conedera, M., Ammann, B. and Lotter, A.F. (2005) Fire ecology north and south of the Alps since the last ice age. *The Holocene* 15, 1214–1226.

Turner, B.L. II, Klepeis, P. and Schneider, L.C. (2003) Three millennia in the southern Yucatán peninsular region: implications for occupancy, use and carrying capacity. In: Gómez-Pompa, A., Allen, M., Fedick, S.L. and Jiménez-Osornio, J. (eds) *The Lowland Maya Area: Three Millennia at the Human–Wildland Interface*. Halworth Press, New York, pp. 361–387.

Turvey, S.T. (ed.) (2009) *Holocene Extinctions*. Oxford University Press, Oxford, UK.

van der Knaap, W.O. and van Leeuwen, J.F.N. (1995) Holocene vegetation succession and degradation as responses to climatic change and human activity in the Serra de Estrela, Portugal. *Review of Palaeobotany and Palynology* 89, 153–211.

Virah-Sawmy, M., Willis, K.J. and Gillson, L. (2010) Evidence for drought and forest declines during the recent megafaunal extinctions in Madagascar. *Journal of Biogeography* 37, 506–519.

Wacnik, A., Goslar, T. and Czernik, J. (2012) Vegetation changes caused by agricultural societies in the Great Mazurian Lake District. *Acta Palaeobotanica* 52, 59–104.

Wang, Z., Li, M., Zhang, R., Zhuang, C., Liu, Y., Saito, Y., Xie, J. and Zhao, B. (2011) Impacts of human activity on the late-Holocene development of the subaqueous Yangtze delta, China, as shown by magnetic properties and sediment accumulation rates. *The Holocene* 21, 393–407.

Watkins, R., Scourse, J.D. and Allen, J.R.M. (2007) The Holocene vegetation history of the Arfon Platform, North Wales, UK. *Boreas* 36, 170–181.

Wen, H.R. and Wen, R.S. (2006) [The forest change in Ningxia during different historical periods]. In: Wen, H.R. (ed.) [*The Changes in the Plants and Animals in China during Different Historical Periods*]. Chongqing Press, Chongqing (in Chinese).

Weng, C., Bush, M.B. and Chepstow-Lusty, A.J. (2004) Holocene changes of Andean alder (*Alnus acuminata*) in highland Ecuador and Peru. *Journal of Quaternary Science* 19, 685–691.

Wick, L. and Tinner, W. (1997) Vegetation changes and timberline fluctuations in the Central Alps as indicators of Holocene climatic oscillations. *Arctic and Alpine Research* 29, 445–458.

Wick, L., van Leeuwen, J.F.N., van der Knaap, W.O. and Lotter, A.F. (2003) Holocene vegetation development in the catchment of Sägistalsee (1935 m asl), a small lake in the Swiss Alps. *Journal of Paleolimnology* 30, 261–272.

Xu, J. (1998) Naturally and anthropogenically accelerated sedimentation in the Lower Yellow River, China, over the past 13,000 years. *Geografiska Annaler: Series A, Physical Geography* 80, 67–78.

Yang, X.D., Ji, S., Jones, R.T., Wang, S.M., Tong, G.B. and Zhang, Z.K. (2005) Pollen evidence of early human activities in Eirhai basin, Yunnan Province. *Chinese Science Bulletin* 50, 568–576.

Yasuda, Y., Kitagawa, H. and Nakagawa, T. (2000) The earliest record of major anthropogenic deforestation in the Ghab Valley, northwest Syria: a palynological study. *Quaternary International* 73/74, 127–136.

Yasuda, Y., Fujiki, T., Nasu, H., Kato, M., Morita, Y., Mori, Y., Kanehara, M., Toyama, S., Yano, A., Okuno, M., Jiejun, H., Ishihara, S. *et al.* (2004) Environmental archaeology at the Chengtoushan site, Hunan Province, China, and implications for environmental change and the rise and fall of the Yangtze River civilization. *Quaternary International* 123–125, 149–158.

Yi, S.H., Saito, Y., Oshima, H., Zhou, Y.Q. and Wei, H.L. (2003a) Holocene environmental history inferred from pollen assemblages in the Huanghe (Yellow River) delta, China: Climatic change and human impact. *Quaternary Science Reviews* 22, 609–628.

Yi, S.H., Saito, Y., Zhao, Q.H. and Wang, P.X. (2003b) Vegetation and climate changes in the Changjiang (Yangtze River) Delta, China, during the past 13,000 years inferred from pollen records. *Quaternary Science Reviews* 22, 1501–1519.

Zanchetta, G., Sulpizio, R., Roberts, N., Cioni, R., Eastwood, W.J., Siani, G., Caron, B., Paterne, M. and Santacroce, R. (2011) Tephrostratigraphy, chronology and climatic events of the Mediterranean basin during the Holocene: an overview. *The Holocene* 21, 33–52.

Zhang, K., Zhao, Y., Zhou, A. and Sun, H. (2010) Late Holocene vegetation dynamic and human activities reconstructed from lake records in western Loess Plateau, China. *Quaternary International* 227, 38–45.

Zhao, Y., Chen, F., Zhou, A., Yu, Z. and Zhang, K. (2010) Vegetation history, climate change and human activities over the last 6200 years on the Liupan Mountains in the southwestern Loess Plateau in central China. *Palaeogeography, Palaeoclimatology, Palaeoecology* 293, 197–205.

3 Contemporary Drivers of Habitat Fragmentation

William F. Laurance*
James Cook University, Queensland, Australia

Abstract

Habitat fragmentation is caused by the clearing and subdivision of formerly continuous vegetation. This chapter describes some geographical patterns of vegetation clearing, the major drivers of this clearing and recent trends in these drivers. Industrial drivers of habitat clearing, including large-scale farming, ranching and tree plantations, have grown in importance in recent decades. In addition, logging, mining, fossil fuel and infrastructure projects are creating an economic impetus for road building in many frontier areas, which also promotes habitat loss. In general, old-growth forests are declining rapidly across the planet and are being replaced by fragmented, secondary and logged forests and plantation monocultures. Forest regeneration may partially mitigate habitat fragmentation in some regions, but rapidly expanding plantations of non-native tree species typically have limited benefits for biodiversity.

Introduction

Humankind has dramatically transformed much of the Earth's surface and its natural biomes and ecosystems. This process has been ongoing for millennia (see Chapter 2, this volume) but has accelerated sharply in the last two centuries, especially in recent decades (e.g. Laurance *et al.*, 2001; Foley *et al.*, 2005; Koh *et al.*, 2011). The fragmentation of many formerly continuous habitats is a frequent result.

Today, the loss and alteration of native habitats continues apace. Many native ecosystems are being progressively razed, bulldozed and felled, until only small remnants of their original vegetation survive. Forests have fared especially poorly, with the global area of forests reduced by roughly half over the past three centuries. Twenty-five nations have now lost virtually all of their forest cover, and another 29 more than nine-tenths of their forest (Millennium Ecosystem Assessment, 2005). Although rates of loss have declined somewhat, tropical forests are still disappearing at an average rate of over 90,000 km^2 per year (Forest Resources Assessment, 2010), equivalent to 35 football fields a minute. Other ecosystems are less imperiled, and a few are even recovering somewhat following centuries of past overexploitation (Fig. 3.1).

This chapter provides an overview of contemporary habitat loss, the principal driver of habitat fragmentation. The patterns of habitat destruction are briefly evaluated geographically and contrasted among different

*bill.laurance@jcu.edu.au

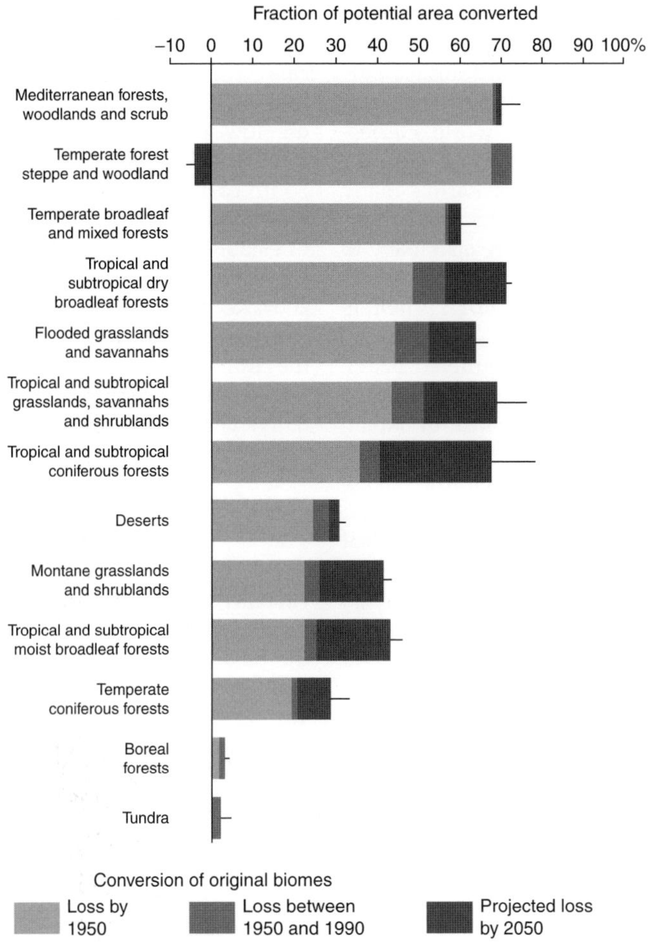

Fig. 3.1. Estimated losses of major terrestrial biomes prior to 1950 and from 1950 to 1990, with projected losses up to 2050. (From Millennium Ecosystem Assessment, 2005.)

biomes and ecosystems. Some of the ultimate and proximate factors that drive habitat loss, and how they are changing over time, are also considered.

Geography of Habitat Loss

Even with the rising spectre of climate change and other environmental perils, habitat disruption is still considered the most important driver of species extinction worldwide (Laurance, 1999; Pimm and Raven, 2000). Rather than being destroyed entirely, most habitats are reduced in extent and simultaneously fragmented, with small remnants of the original habitat persisting like islands in a sea of modified land.

Certain regions of the Earth are far more affected by habitat destruction than others. Among the most imperiled are the so-called 'biodiversity hotspots', which contain high species endemism (at least 1500 locally endemic plant species) and which have lost at least 70% of their original native vegetation (Myers *et al.*, 2000). Over half of the 35 recognized terrestrial hotspots are located in the tropics (Mittermeier *et al.*, 2005; Williams *et al.*, 2011). The Atlantic forests of Brazil and rainforests of West Africa, both of which have been severely reduced and degraded, are examples. Despite encompassing just a small fraction of the Earth's land surface, surviving habitats in biodiversity hotspots might sustain over half of the world's terrestrial species (Myers *et al.*, 2000; Mittermeier *et al.*, 2005). A majority of the Earth's many undescribed species, including large numbers of restricted endemics, are also likely to occur in biodiversity hotspots (Joppa *et al.*, 2011; Laurance and Edwards, 2011; Scheffers *et al.*, 2012).

Many islands have also suffered heavy habitat losses. For instance, most native habitats have already been lost in Japan, New Zealand, Madagascar, the Philippines and Java (World Resources Institute, 2003). Other islands, such as Borneo, Sumatra and New Guinea, still retain expanses of original habitat but are losing it at alarming rates (Curran *et al.*, 2004; Laurance, 2007a; Koh *et al.*, 2011). This is alarming because any island systems, such as Madagascar, New Zealand and the Sunda Shelf Islands of South-east Asia, are also important biodiversity hotspots.

Most areas of high human population density have suffered heavy habitat disruption. Such areas include much of Europe, eastern North America, South and South-east Asia, the Middle East, West Africa, Central America and the Caribbean region, among others. Biodiversity hotspots often coincide with areas of high population density (Cincotta *et al.*, 2000), and the hotspots that are losing remaining habitat the fastest are also those most plagued by poor standards of living and low human development (Jha and Bawa, 2006). Human populations are dense in many coastal areas, which have experienced considerable losses of both terrestrial habitats and nearby coral reefs. Coastal zones in Asia, northern South America, the Caribbean, Europe and eastern North America have all suffered heavy habitat losses (Millennium Ecosystem Assessment, 2005). The intense vulnerability of many islands might be largely attributable to their tendency to have high human population densities, although there does not appear to be any systematic test of this idea.

Finally, habitat destruction can occur swiftly in areas with limited human densities but rapidly expanding agriculture. Large expanses of the Amazon, for example, are currently being cleared for industrial-scale cattle ranching and soy farming (Laurance *et al.*, 2001; Fearnside *et al.*, 2012), despite having low population densities. Likewise, in some relatively sparsely populated areas of South-east Asia, such as Borneo, Sumatra and New Guinea, forests are rapidly being felled to establish oil-palm, rubber and pulpwood plantations (Laurance, 2007a; Ziegler *et al.*, 2009; Koh *et al.*, 2011). Older agricultural frontiers, such as those in Europe, eastern China, the Indian subcontinent, and eastern and mid-western North America, often have little native vegetation remaining (Fig. 3.2).

Changing Drivers of Habitat Disruption

Globally, agriculture is the biggest cause of habitat destruction (Foley *et al.*, 2005). Over the past three centuries, the extent of cropland globally (Fig. 3.2) has risen sharply, from around 2.7 to 15 million km^2 (Turner *et al.*, 1990). Pastures and grazing lands are even more extensive, spanning around 34 million km^2 by the mid-1990s (Wood *et al.*, 2000). In total, farmlands now encompass an area the size of South America (Fig. 3.2), and grazing and pasture lands an area the size of Africa (Foley *et al.*, 2005).

We are currently experiencing a dramatic increase in the global footprint of agriculture. For instance, more land was converted to cropland from 1950 to 1980 than from 1700 to 1850 (Millennium Ecosystem Assessment, 2005). This acceleration will surely continue this century, given the rapid increases in population and food demand. Based on current trends in farming practices, meeting projected food demand will require about 1 billion ha of additional farming and grazing land by 2050 (Tilman *et al.*, 2011).

In developing nations, where most habitat loss is currently occurring, the drivers of environmental change have shifted fundamentally in recent decades. Instead of being caused mostly by small-scale farmers and rural inhabitants, habitat loss is now substantially driven by globalization and industrial activities (Butler and Laurance, 2008;

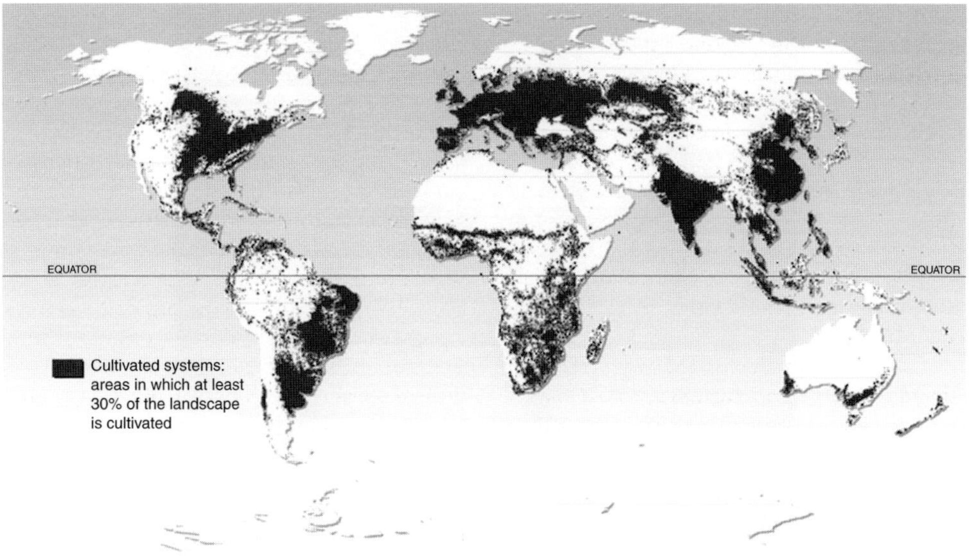

Fig. 3.2. Extent of land area cultivated globally by the year 2000. (From Millennium Ecosystem Assessment, 2005.)

Rudel *et al.*, 2009). These shifting drivers of deforestation are especially evident in the tropics (Fig. 3.3). Prior to the late 1980s, tropical deforestation was largely associated with rapid human population growth in developing nations, in concert with government policies for rural development, such as agricultural loans, tax incentives and road construction (Rudel, 2005). Such initiatives, especially apparent in nations such as Brazil and Indonesia, promoted large influxes of colonists into frontier areas and often caused dramatic forest losses (Fearnside, 1997).

More recently, however, the impacts of rural peoples on tropical forests seem to be stabilizing. Although many tropical nations still have considerable population growth, strong urbanization trends (except in sub-Saharan Africa) mean that rural populations are growing more slowly, and are even declining in some areas (Wright and Muller-Landau, 2006). The popularity of large-scale frontier-colonization programmes has also waned. If such trends continue, they could begin to alleviate some pressures on forests from small-scale farming, hunting and fuelwood gathering (Wright and Muller-Landau, 2006; but see Laurance, 2007b).

At the same time, globalized financial markets and a worldwide commodity boom are creating a highly attractive environment for the private sector. Under these conditions, large-scale agriculture – crops, livestock and tree plantations – by corporations and wealthy landowners is increasingly emerging as the largest direct cause of tropical deforestation (Butler and Laurance, 2008; Rudel *et al.*, 2009). Surging demand for grains and edible oils (Koh and Lee, 2012), driven by the global thirst for biofuels and rising standards of living in developing countries, is also spurring this trend. In Brazilian Amazonia, for instance, large-scale ranching has exploded in recent years, with the number of cattle more than tripling, from 22 to 74 million head, since 1990 (Smeraldi and May, 2008). Industrial soy farming has also grown dramatically (Fearnside, 2001; Barona *et al.*, 2010).

Other industrial activities, including logging, mining, fossil fuel and infrastructure developments, are also playing a major but indirect role in forest destruction (Laurance *et al.*, 2001; Asner *et al.*, 2006; Finer *et al.*, 2008; Finer and Jenkins, 2012; see Chapter 10, this volume). These provide a key economic impetus for forest road building, which in turn

Fig. 3.3. Changing drivers of deforestation: small-scale cultivators (a) versus industrial road construction (b) in Gabon, Central Africa. (Photographs by W.F. Laurance.)

facilitates movements of colonists, hunters and miners into frontier areas, often leading to serious forest disruption and cycles of land speculation (Laurance *et al.*, 2009). There is a dire need to plan road expansions more carefully and proactively, especially in remaining wilderness areas (Laurance and Balmford, 2013).

Land-use Intensification and Abandonment

Globally, the net rate of conversion of native habitats has finally begun to slow, because readily available arable land is declining and because, in temperate and boreal regions, some ecosystems are recovering somewhat. Forest cover is now increasing in eastern and western North America, Alaska, western and northern Europe, eastern China and Japan (Matthews *et al.*, 2000; Millennium Ecosystem Assessment, 2005). During the 1990s, forest cover rose by around 29,000 km^2 annually in the temperate and boreal zones, although roughly 40% of this was for forest plantations mostly comprising non-native tree species (Millennium Ecosystem Assessment, 2005). Despite partial recovery of forest cover in some regions, conversion rates for many ecosystems, such as tropical and subtropical forests, the South American Pantanal and Cerrado savannah woodlands, and the Miombo and Guinea savannah woodlands of Africa, remain high (Sodhi *et al.*, 2004; Klink and Machado, 2005; Mayaux *et al.*, 2005; Forest Resources Assessment, 2010).

Because the amount of arable land is diminishing while demands for food and biofuels are still rising markedly, agriculture is becoming increasingly intensified in much of the world. Within many agricultural regions, a greater fraction of the available land is being cultivated, the intensity of cultivation is increasing and fallow lengths are decreasing (Millennium Ecosystem Assessment, 2005). Cultivated systems (where over 30% of the landscape is in croplands, shifting cultivation, confined-livestock production or freshwater aquaculture) spanned 24% of the global land surface by the year 2000 (Fig. 3.2). This trend is not universal, and productivity is actually falling in some regions, but agricultural intensification will surely continue in many areas this century (Foley *et al.*, 2011).

Thus, vast expanses of the Earth have already been altered by human activities. Old-growth forests have diminished greatly in extent in many regions, especially in temperate zones; for instance, at least 94% of temperate broadleaf forests have been disturbed by farming and logging (Primack, 2006). Other ecosystems, such as New and Old World coniferous forests and Australian ash forests, are rapidly being converted from old-growth to semi-natural production forests with few large trees and a simplified stand structure and species composition (Lindenmayer *et al.*, 2012). Forest cover is increasing in parts of the temperate and boreal zones, but the new forests are secondary and differ from old-growth forests in species composition, structure and carbon storage (Franklin *et al.*, 1981; Lindenmayer *et al.*, 2000). Yet other ecosystems, particularly in the tropics, are rapidly being destroyed and degraded.

The large-scale transformations of land cover described here consider only habitat loss per se. Of the surviving habitat, much is being degraded in various ways, such as by habitat fragmentation, edge effects, selective logging, pollution, overhunting, altered fire regimes and climate change. These forms of environmental degradation can interact synergistically with habitat fragmentation (Laurance and Cochrane, 2001), leading to even greater impacts on biodiversity and other environmental values (Brook *et al.*, 2008; Laurance and Useche, 2009). Predicting the consequences of such potentially complex interactions is a major challenge for conservation biologists.

Acknowledgements

I thank an anonymous referee and the book editors for commenting on this chapter, and the Australian Research Council for support.

References

Asner, G.P., Broadbent, E., Oliveira, P., Keller, M., Knapp, D. and Silva, J. (2006) Condition and fate of logged forests in the Amazon. *Proceedings of the National Academy of Sciences USA* 103, 12947–12950.

Barona, E., Ramankutty, N., Hyman, G. and Coomes, O. (2010) The role of pasture and soybean in deforestation of the Brazilian Amazon. *Environmental Research Letters* 5, 024002.

Brook, B.W., Sodhi, N.S. and Bradshaw, C.J.A. (2008) Synergies among extinction drivers under global change. *Trends in Ecology and Evolution* 23, 453–460.

Butler, R.A. and Laurance, W.F. (2008) New strategies for conserving tropical forests. *Trends in Ecology and Evolution* 23, 469–472.

Cincotta, R.P., Wisnewski, J. and Engelman, R. (2000) Human population in the biodiversity hotspots. *Nature* 404, 990–992.

Curran, L.M., Trigg, S., McDonald, A., Astiani, D., Hardiono, Y., Siregar, P., Caniago, I. and Kasischke, E. (2004) Lowland forest loss in protected areas of Indonesian Borneo. *Science* 303, 1000–1003.

Fearnside, P.M. (1997) Transmigration in Indonesia: lessons from its environmental and social impacts. *Environmental Management* 21, 553–570.

Fearnside, P.M. (2001) Soybean cultivation as a threat to the environment in Brazil. *Environmental Conservation* 28, 23–38.

Fearnside, P.M., Laurance, W.F., Cochrane, M.A., Bergen, S., Delamonica, P., Barber, C., D'Angelo, S. and Fernandes, T. (2012) A futuro da Amazônia: modelos para prever as conseqüências da infraestrutura futura nos planos plurianuais. *Novos Cadernos NAEA* 15, 25–52.

Finer, M. and Jenkins, C.N. (2012) Proliferation of hydroelectric dams in the Andean Amazon and implications for Andes–Amazon connectivity. *PLoS One* 7, e35126.

Finer, M., Jenkins, C., Pimm, S., Kean, B. and Rossi, C. (2008) Oil and gas projects in the western Amazon: threats to wilderness, biodiversity, and indigenous peoples. *PLoS One* 3, e2932.

Foley, J.A., DeFries, R., Asner, G.P., Barford, C., Bonan, G., Carpenter, S.R., Chapin, F.S., Coe, M.T., Dailey, G.C., Gibbs, H.K., Helkowski, J.H., Holloway, T. *et al.* (2005) Global consequences of land use. *Science* 309, 570–574.

Foley, J.A., Ramankutty, N., Brauman, K.A., Cassidy, E.S., Gerber, J.S., Johnston, M., Mueller, N.D., O'Connell, C., Ray, D.K., West, P.C., Balzer, C., Bennett, E.M. *et al.* (2011) Solutions for a cultivated planet. *Nature* 478, 337–342.

Forest Resources Assessment (2010) *Global Forest Resources Assessment 2010*. United Nations Food and Agricultural Organization, Rome, Italy.

Franklin, J.F., Cromack, K., Denison, W., McKee, A., Maser, C., Sedell, J., Swanson, F. and Juday, G. (1981) *Ecological Characteristics of Old-growth Douglas-Fir Forests*. US Forest Service, Pacific Northwest Research Station, Portland, Oregon, USA.

Jha, S. and Bawa, K.S. (2006) Population growth, human development, and deforestation in biodiversity hotspots. *Conservation Biology* 20, 906–912.

Joppa, L.N., Roberts, D.L., Myers, N. and Pimm, S.L. (2011) Biodiversity hotspots house most undiscovered plant species. *Proceedings of the National Academy of Sciences USA* 108, 13171–13176.

Klink, C.A. and Machado, R.B. (2005) Conservation of the Brazilian cerrado. *Conservation Biology* 19, 707–713.

Koh, L.P. and Lee, T.M. (2012) Sensible consumerism for environmental sustainability. *Biological Conservation* 151, 3–6.

Koh, L.P., Miettinen, J., Liew, S.C. and Ghazoul, J. (2011) Remotely sensed evidence of tropical peatland conversion to oil palm. *Proceedings of the National Academy of Sciences USA* 108, 5127–5132.

Laurance, W.F. (1999) Reflections on the tropical deforestation crisis. *Biological Conservation* 91, 109–117.

Laurance, W.F. (2007a) Forest destruction in tropical Asia. *Current Science* 93, 1544–1550.

Laurance, W.F. (2007b) Have we overstated the tropical biodiversity crisis? *Trends in Ecology and Evolution* 22, 65–70.

Laurance, W.F. and Balmford, A. (2013) A global map for road building. *Nature* 495, 308–309.

Laurance, W.F. and Cochrane, M.A. (2001) Synergistic effects in fragmented landscapes. *Conservation Biology* 15, 1488–1489.

Laurance, W.F. and Edwards, D.P. (2011) The search for hidden biodiversity. *Proceedings of the National Academy of Sciences USA* 108, 12971–12972.

Laurance, W.F. and Useche, D.C. (2009) Environmental synergisms and extinctions of tropical species. *Conservation Biology* 23, 1427–1437.

Laurance, W.F., Cochrane, M.A., Bergen, S., Fearnside, P.M., Delamonica, P., Barber, C., D'Angelo, S. and Fernandes, T. (2001) The future of the Brazilian Amazon. *Science* 291, 438–439.

Laurance, W.F., Goosem, M. and Laurance, S.G. (2009) Impacts of roads and linear clearings on tropical forests. *Trends in Ecology and Evolution* 24, 659–669.

Lindenmayer, D.B., Cunningham, R.B., Donnelly, C.F. and Franklin, J.F. (2000) Structural features of old growth Australian montane ash forests. *Forest Ecology and Management* 134, 189–204.

Lindenmayer, D.B., Laurance, W.F. and Franklin, J.F. (2012) Global decline in large old trees. *Science* 338, 1305–1306.

Matthews, E., Rohweder, M., Payne, R. and Murray, S. (2000) *Pilot Analysis of Global Ecosystems: Forest Ecosystems*. World Resources Institute, Washington, DC.

Mayaux, P., Holmgren, P., Achard, F., Eva, H., Stibig, H.-J. and Branthomme, A. (2005) Tropical forest cover change in the 1990s and options for future monitoring. *Philosophical Transactions of The Royal Society B: Biological Sciences* 360, 373–384.

Millennium Ecosystem Assessment (2005) *Ecosystems and Human Well-Being: Synthesis*. Island Press, Washington, DC.

Mittermeier, R.A., Robles, P., Hoffman, M., Pilgrim, J., Brooks, T., Mittermeier, C.G., Lamoreux, J. and da Fonseca, G.A.B. (2005) *Hotspots Revisited: Earth's Biologically Richest and Most Endangered Terrestrial Ecosystems*. University of Chicago Press, Chicago, Illinois.

Myers, N., Mittermeier, R., Mittermeier, C., Fonseca, G. and Kent, J. (2000) Biodiversity hotspots for conservation priorities. *Nature* 403, 853–858.

Pimm, S.L. and Raven, P. (2000) Biodiversity: extinction by numbers. *Nature* 403, 843–845.

Primack, R.B. (2006) *Essentials of Conservation Biology*, 4th edn. Sinauer Associates, Sunderland, Massachusetts.

Rudel, T.K. (2005) Changing agents of deforestation: from state-initiated to enterprise driven processes, 1970–2000. *Land Use Policy* 24, 35–41.

Rudel, T.K., DeFries, R., Asner, G.P. and Laurance, W.F. (2009) Changing drivers of tropical deforestation create new challenges and opportunities for conservation. *Conservation Biology* 23, 1396–1405.

Scheffers, B.R., Joppa, L., Pimm, S.L. and Laurance, W.F. (2012) What we know and don't know about Earth's missing biodiversity. *Trends in Ecology and Evolution* 27, 501–510.

Smeraldi, R. and May, P.H. (2008) *The Cattle Realm: a New Phase in the Livestock Colonization of Brazilian Amazonia*. Amigos da Terra, Amazônia Brasileira, São Paulo, Brazil.

Sodhi, N.S., Koh, L.P., Brook, B.W. and Ng, P. (2004) Southeast Asian biodiversity: an impending catastrophe. *Trends in Ecology and Evolution* 19, 654–660.

Tilman, D., Balzer, C., Hill, J. and Befort, B.L. (2011) Global food demand and the sustainable intensification of agriculture. *Proceedings of the National Academy of Sciences USA* 108, 20260–20264.

Turner, B.L., Clark, W.C., Kates, R., Richards, J., Mathews, J. and Meyer, W. (eds) (1990) *The Earth as Transformed by Human Action: Global and Regional Change in the Biosphere over the Past 300 Years*. Cambridge University Press, Cambridge, UK.

Williams, K.J., Ford, A., Rosauer, D., de Silva, N., Mittermeier, R., Bruce, C., Larsen, F.W. and Margules, C. (2011) The forests of East Australia: The 35th biodiversity hotspot. In: Zachos, F.E. and Habel, J.C. (eds) *Biodiversity Hotspots: Distribution and Protection of Priority Conservation Areas*. Springer, Berlin, pp. 295–310.

Wood, S., Sebastian, K. and Scherr, S.J. (2000) *Pilot Analysis of Global Ecosystems: Agroecosystems*. World Resources Institute, Washington, DC.

World Resources Institute (2003) *World Resources 2002–2004: Decisions for the Earth: Balance, Voice, and Power*. World Resources Institute, Washington, DC.

Wright, S.J. and Muller-Landau, H.C. (2006) The future of tropical forest species. *Biotropica* 38, 287–301.

Ziegler, A.D., Fox, J.M. and Xu, J. (2009) The rubber juggernaut. *Science* 324, 1024–1025.

4 Forest Fragmentation and Biodiversity Conservation in Human-dominated Landscapes

Timm F. Döbert,[1,2] Bruce L. Webber,[1,2] Andrew D. Barnes,[3] Katharine J.M. Dickinson[4] and Raphael K. Didham[1,2]*

[1]*University of Western Australia;* [2]*CSIRO Ecosystem Sciences, Perth, Australia;* [3]*University of Göttingen, Germany;* [4]*University of Otago, New Zealand*

Abstract

The future of forest biodiversity is increasingly dependent on the ability of species to persist in isolated forest remnants embedded within a mosaic of human land uses. We review the impacts of forest loss, habitat degradation and alteration of the spatial structure of the landscape on connectivity and population persistence in fragmented landscapes. Conservation management is just beginning to come to grips with the challenges faced in moving away from the traditional patch-focused approach of conservation in gazetted nature reserves towards a landscape-focused approach of conserving biodiversity in a mosaic that includes managed and semi-natural habitats. Contentious debate has centred on the ability of degraded forest systems to sustainably support forest specialist species, and the utility of traditional species–area approaches to predict biodiversity loss relationships without appropriately accounting for varying species responses to landscape matrix quality and edge effects. While there is no doubt that some area-sensitive and disturbance-sensitive species require large areas of primary forest to ensure population persistence, there is an emerging consensus that the maintenance of biodiversity will depend as much, if not more, on the extent, magnitude and spatial structuring of landscape processes within the degraded matrix surrounding primary forest remnants. Accordingly, biodiversity conservation must refocus more on the interaction between patch and landscape processes than on patch processes per se. This will demand more effective discrimination of the relative importance of total habitat loss, declining habitat quality and altered spatial structuring of suitable habitat as mechanistic drivers of biodiversity loss in fragmented forest landscapes.

Introduction

It is now rare, if not impossible, in most regions of the world to find forested landscapes unmodified by human influence. The primeval forests that once covered 6 billion hectares of the Earth's land surface (Bryant *et al.*, 1997) have suffered widespread clearance for agriculture and human settlement, and much of what remains has been subject to

*raphael.didham@uwa.edu.au

intensive management for the extraction of natural resources (Ellis *et al.*, 2010). Today, the majority of old-growth forest remnants are small and scattered and are of extremely high conservation value within heavily fragmented landscapes. Effective biodiversity conservation in this context demands a detailed understanding of threatening processes that operate not only at the fragment level but also at the surrounding landscape level within which fragments are embedded (Ewers and Didham, 2006a; Lindenmayer and Fischer, 2006).

The importance of a diverse set of 'landscape-context effects' on isolated forest fragments has been widely appreciated since at least the early 1980s (e.g. Janzen, 1983) and has been reviewed extensively in the ecological literature since the development of a fully fledged landscape perspective on habitat fragmentation in the 1990s (e.g. Ries *et al.*, 2004; Tscharntke *et al.*, 2005a; Ewers and Didham, 2006a; Kupfer *et al.*, 2006; Collinge, 2009; Tscharntke *et al.*, 2012). However, there is still much contention among ecologists about the relative importance of patch-level versus landscape-level processes influencing fragment communities (Didham *et al.*, 2012b), and debate remains about the most appropriate conceptual approach to partitioning their relative effects in a hierarchical landscape framework (Box 4.1).

Moreover, management approaches to biodiversity conservation in complex landscape mosaics have lagged much further behind both our empirical and theoretical understanding of species responses to landscape change. Although there has been some management uptake (at least in a few regions) of concepts such as 'corridors improve connectivity' (Beger *et al.*, 2010) and 'the matrix matters' for populations in isolated fragments (Brady *et al.*, 2009; see Chapter 7, this volume), conservation prioritization is still, by and large, based on protection of patch-level assets rather than landscape-level processes (Wiens, 2009). As the rate of attrition of old-growth habitat continues to increase unabated (e.g. Laurance, 2007; FAO, 2010), it appears to be only a matter of time before conservation practice must make the transition from patch-level to landscape-level objectives, simply because there will be insufficient 'natural' habitat left within which to conserve biodiversity. Already, there has been a palpable shift in academic thinking about the optimal approach to biodiversity conservation in human-dominated landscapes, with the strong 'wilderness conservation' ethos of the 20th century rapidly giving way to a 'biodiversity on degraded lands' ethos over the last decade (Box 4.2). Whether this 'new conservation paradigm' (Chazdon *et al.*, 2009) will enhance or devalue the importance of old-growth remnants in the landscape remains to be seen (Box 4.2), with some authors even calling for the scrapping of 'under-performing' national parks in order to reallocate funding to other conservation management actions (Fuller *et al.*, 2010).

Here, we argue that the layering, or integration, of new conservation efforts on degraded lands on to existing conservation efforts in discrete, gazetted nature reserves (Mittermeier *et al.*, 2003; Tscharntke *et al.*, 2005a; Harvey *et al.*, 2008; Gardner *et al.*, 2009) will do more to enhance biodiversity conservation in the landscape than either approach alone. In this chapter, we briefly review the ecological principles underpinning the effects of altered spatial structuring of the landscape on remnant communities, focusing on the influential role played by landscape matrix processes in determining patch-level outcomes. We discuss six key hypotheses for the ways in which these landscape-context effects can moderate biodiversity patterns and processes in fragmented landscapes. We then consider how a greater understanding of these principles might improve the allocation of conservation effort under different scenarios of landscape change, species sensitivity to disturbance and the shifting socio-economic drivers of commodity production over conservation outcomes. In some senses, when viewed through the lens of species resilience to habitat degradation within the matrix, the ecological implications of forest fragmentation for biodiversity conservation might well be less severe than previously considered in the past. However, it is likely that this apparent resilience to human disturbance is only maintained while sufficient old-growth

Box 4.1. Shifting ecological paradigms in the study of habitat fragmentation

One widely adopted definition of habitat fragmentation is 'the process by which habitat loss results in the division of large, continuous habitats into a greater number of smaller patches of lower total area, isolated from each other by a matrix of dissimilar habitats' (modified from Wilcove *et al.*, 1986; Ewers and Didham, 2006a). This definition nicely reflects the fact that most fragmentation studies today either implicitly or explicitly link both patch-scale and landscape-scale perspectives on the ecological consequences of land-use change for organisms living in a network of remnant patches surrounded by a mosaic of modified or novel land-use types (Didham, 2010). However, this has not always been the case. Historically, habitat fragmentation theory had its roots in the simplistic stochastic model of island biogeography theory (IBT; MacArthur and Wilson, 1967), and was undoubtedly much more strongly patch-focused than it is today. For instance, in its strict form, IBT considered just patch area and isolation, and incorporated no external influence from the inhospitable matrix and no internal patch dynamics (Didham, 2010). Despite the fact that this is obviously a poor caricature of reality (Laurance, 2008), some authors have continued to conflate 'fragmentation' models in general with an idealized 'IBT' model (Fischer and Lindenmayer, 2006; Didham *et al.*, 2012b).

Perhaps the most influential factor in cementing the landscape-scale context to fragmentation was the seminal review by Fahrig (2003). She formalized 'fragmentation per se' as an expression of the spatial pattern of arrangement of habitat patches at a given level of habitat loss, rather than as an expression of the overall temporal process of changing landscape structure. The rationale was to tackle the ubiquitous collinearity between habitat loss and altered spatial arrangement of habitat (Andrén, 1994; Fahrig, 1997), and this has since become one of the defining paradigm shifts in the study of habitat fragmentation (Fahrig, 2003; Collinge, 2009; Didham *et al.*, 2012b). As Didham *et al.* (2012b) put it, it has become de rigeur to control for habitat loss effects first before interpreting the remaining variance attributable to altered spatial arrangement of habitat. Unfortunately, however, this has fostered the impression that habitat loss and habitat fragmentation can legitimately be considered as 'independent' of one another, even though the two are highly collinear in real landscapes. What is more, the generalized conclusion from this perspective has been that fragmentation (at the landscape scale) has little or no effect on species richness after controlling for habitat loss, which stands at odds with the wealth of evidence for the ecological impact of fragmentation-related variables such as fragment area, edge effects and matrix contrast on communities at the patch scale (Didham, 2010; Didham *et al.*, 2012b). Instead, Didham *et al.* (2012b) made the argument that what is urgently needed is a formal conceptual model that better reflects the hierarchical causal dependence between habitat loss and habitat fragmentation across patch and landscape scales, and at the same time can statistically partition the indirect contributions of habitat loss and fragmentation to the collinear portion of variance among them (e.g. a structural equation modelling approach). It is believed that such a conceptual shift towards hierarchical causal models, incorporating interdependence in the effects of multiple habitat variables across multiple spatial scales, would provide a much greater mechanistic understanding of the ecological consequences of habitat fragmentation (Didham *et al.*, 2012b). Undoubtedly, there will be fertile ground for further re-evaluation of the key paradigms underpinning the study of habitat fragmentation in the future.

source populations remain as refuges in the landscape. Consequently, we conclude with the message that it will always remain a priority to focus conservation efforts on limiting and mitigating the effects of forest clearing and fragmentation on old-growth remnants, even when there are substantial conservation opportunities in the surrounding matrix.

Principles of Spatial Arrangement Effects in a Fragmented Landscape

Organisms living in fragmented landscapes are subject to a wide range of factors operating at different spatial scales, from the local (within-patch) scale, to the fragment scale, to the surrounding landscape scale. Although

Box 4.2. Shifting management approaches to conserving forest biodiversity

Although human impacts on biodiversity can be comparatively straightforward to describe in ecological terms, their implications for conservation management depend heavily on a set of subjective value judgements about which components of biodiversity are important to conserve, and on how society should weigh up the relative importance of environmental versus cultural versus economic outcomes in a given region. In this sense, the 'conservation' component of conservation management might be considered to be just as much of a societal commodity as more typical production commodities (such as timber) that are managed in forested landscapes. With the massive changes that have occurred over the last 30 years in terms of human population pressure, increasing resource demands and declining environmental quality, it is perhaps not surprising that changing societal values are forcing a dramatic shift in the paradigms underpinning conservation management (Didham, 2011).

The strong 'wilderness conservation' ethos of the 1980s and 1990s is rapidly giving way to a 'biodiversity on degraded lands' ethos (Chazdon *et al.*, 2009), with the realization that gazetted nature reserves alone will be insufficient to conserve even a fraction of global biodiversity. Numerous interlinked research agendas now champion the conservation of biodiversity in human-dominated landscapes (Gardner *et al.*, 2009), and focus much more heavily than ever before on 'conservation for the people' (Kareiva *et al.*, 2007) and on the ecosystem services that biodiversity can provide to humans (Nicholson *et al.*, 2009) in so-called 'novel ecosystems' (Hobbs *et al.*, 2006). However, the commodification of nature is not without its critics (McCauley, 2006; Didham, 2011). The issue is not so much whether degraded lands can have significant conservation value (they obviously can in some instances) but whether the 'biodiversity on degraded lands' ethos will successfully enhance, rather than destabilize, existing conservation efforts (Didham, 2011). It will be crucial not to overgeneralize the notion that conservation on degraded lands is a viable alternative (rather than a complement) to conservation of natural ecosystems, or to portray the intrinsic ability of species to persist and recover in heavily degraded landscapes as a 'trump card' for biodiversity conservation (Tabarelli, 2010). There will be some ideal scenarios where it is possible to retain high biodiversity at the landscape scale despite a low proportion of remaining old-growth forest, because remnants are embedded within a complex landscape matrix of low-intensity land uses (such as shade-grown coffee in tropical regions; Perfecto and Vandermeer, 2008; see Chapter 7, this volume). However, by far the majority of human-dominated landscapes do not fit this model, and instead they have more dichotomous landscapes of small remaining semi-natural areas within extensive agricultural landscapes of low native biodiversity (Rudel *et al.*, 2009). In these latter cases, reliance on conservation through the biodiversity retained on degraded lands would be ineffective, if not foolhardy. Of course, these scenarios represent polarized extremes of what is really a continuum of approaches to land-sharing versus land-sparing as ways of balancing the trade-offs between production and biodiversity values in managed landscapes (Fischer *et al.*, 2008). Clearly, much greater care is needed in qualifying the conflicting messages that 'human impact degrades biodiversity' yet 'degraded habitats have high biodiversity value' (Didham, 2011). We believe that the conservation of biodiversity on degraded lands should only ever be one component of wider conservation strategies centred on a network of protected areas in which negative human impacts are limited as much as possible.

the patch-scale effects of fragment area and degree of spatial isolation – the key spatial components of the classical island biogeography theory (IBT; Box 4.1) – continue to receive substantial research emphasis (Prugh *et al.*, 2008; Struebig *et al.*, 2008), it is well recognized that they represent just two of at least five major spatial components of fragmentation effects, alongside patch shape effects, edge effects and landscape matrix effects (Ewers and Didham, 2006a; Didham, 2010). It is not our intention to review the separate ecological effects of these variables here (as they have already been well reviewed elsewhere) but rather to highlight key principles of the combined (potentially non-additive) effects of multiple spatial components of fragmentation acting in concert.

First, it is important to recognize that patterns of habitat fragmentation are imposed on to a pre-existing template of natural spatial heterogeneity in topography, soil type, vegetation composition and many other biotic and abiotic factors (Ewers and Didham, 2006a). For instance, heterogeneity in underlying species distributions has been recognized to drive 'sampling effects' in the so-called 'cookie-cutter model' of stochastic species occurrence within recently isolated fragments (Laurance *et al.*, 2011) and has been recognized as an important factor to consider when discriminating anthropogenic edge effects from underlying spatial autocorrelation of ecological responses (Baker and Barmuta, 2006). However, the general importance of these types of heterogeneity issues in the study of habitat fragmentation is not as widely appreciated as it should be. There tends to be an implicit assumption in most studies that ecological responses to the anthropogenic drivers of fragmentation can be interpreted directly, when in fact it might be quite common (if not ubiquitous) for underlying spatial heterogeneity to mask or even exacerbate apparent fragmentation effects. These confounding effects are likely to be particularly severe where there is strong collinearity between the drivers of fragmentation and the underlying environmental gradients structuring communities (Seabloom *et al.*, 2002). For example, there tends to be higher habitat loss and lower average fragment size on fertile soil types at low elevations where the land is most suitable for agriculture (e.g. Ewers *et al.*, 2006). Because these areas might intrinsically support higher biodiversity than infertile soils, the underlying environmental gradient might confound interpretation of patch area effects.

Secondly, none of the spatial variables considered in fragmentation models (i.e. habitat loss, fragment area, isolation, shape complexity, edge effects or matrix context) should strictly be thought of as 'effect mechanisms' in their own right (Didham *et al.*, 2012b). Instead, spatial variables are just proxies for one or more biotic or abiotic mechanisms that actually elicit ecological responses (see examples in Blakely and Didham, 2010; Didham *et al.*, 2012b). Quantifying the proximate variables mediating spatial patterns would go a long way towards overcoming inherent collinearity of driving variables. For instance, small forest fragments are characterized by both small fragment area and high edge influence (among other factors), and therefore it might be difficult to discern whether an ecological response, such as declining population size, is due to the mechanistic effects of small-area processes or near-edge processes. Indeed, in the Atlantic forests of Brazil, Banks-Leite *et al.* (2010) found that bird species had congruent responses to edge and area effects, such that the degree of change in bird species composition between forest edges and interiors was similar to the change in community composition between small and large patches. They concluded that apparent fragment area effects were actually driven by edge sensitivity of certain bird species, not by any mechanistic responses related to patch area per se (Banks-Leite *et al.*, 2010).

Thirdly, and most importantly, the different spatial components of forest fragmentation do not act independently of one another. Because fragmentation effects are inherently hierarchical in nature (Box 4.1), with edges nested within patches, patches within landscapes and landscapes within regions, it is not possible to determine the relative importance of a given variable without understanding the spatial and temporal context within which it is measured. For example, Ewers *et al.* (2007) found that local edge effects did not have a consistent influence on beetle biodiversity or community composition independent of the fragment-level context within which the edges were sampled. Instead, they found evidence of a strong synergistic interaction between habitat area and edge effects, with the strength of edge effects changing exponentially with decreasing fragment area (Ewers *et al.*, 2007). Meanwhile, the effect of fragment area on ecological responses can also be highly dependent on a wide range of other spatial components of fragmentation, such as the degree of spatial isolation from other suitable habitat fragments (Bender and Fahrig, 2005). Area by isolation interactions are particularly well recognized from their central role in the dynamic equilibrium of colonization and extinction probabilities underlying IBT (Krauss *et al.*, 2003) and simple metapopulation models (Hanski, 1999), but there are also many other spatial context effects moderating the emergent effects of patch area in fragmented landscapes (Collins *et al.*, 2009). Similarly, spatial isolation itself can be viewed as a relative construct that depends on the interaction between geographical distance, dispersal potential and matrix context (i.e. the 'effective isolation'; Ricketts, 2001). For example, in a metapopulation of the speckled wood butterfly (*Pararge aegeria* L.) in Belgium, Chardon *et al.* (2003) found that effective isolation was a much more accurate predictor of population connectivity than pure Euclidean distance among habitat patches. Numerous other examples of pairwise and multiway interactions among spatial components of fragmentation now abound in the literature and it is likely that context dependence in ecological responses is ubiquitous across systems.

At the landscape scale, the combination of species responses to all of these interacting factors is often expressed cumulatively as the degree of change in biodiversity or community composition with altered habitat configuration. Interpreting these landscape-level habitat configuration effects is fraught with difficulty when viewed in isolation from their mechanistic local- and patch-scale determinants. It must be remembered that habitat configuration in the landscape is inherently dependent on the spatial and temporal progression of habitat loss, resulting in strong intercorrelation between quantitative fragmentation metrics and the degree of habitat loss in the landscape (e.g. Fig. 3 in Fahrig, 2003). Effective partitioning of this intercorrelated portion of variance is only possible with an appropriate model of causal inference among the different components of amount and distribution of habitat across spatial scales (Didham *et al.*, 2012b). Perhaps the most important failing of habitat fragmentation studies in general has been to ignore the causal dependence of patch processes nested within the landscape mosaic, and simply treating each component of habitat arrangement as an independent variable (Box 4.1).

A hierarchical causal model (Fig. 4.1) would allow a formal distinction to be made between the ultimate versus proximate effects of spatial variables and would provide a working platform for testing (rather than assuming) whether individual variables are independent of one another (Didham *et al.*, 2012b). This, in turn, would allow the field to focus on a new generation of more relevant and interesting fragmentation questions revolving around partitioning the relative effects of ultimate versus proximate or direct versus indirect effects of multiple drivers of landscape change.

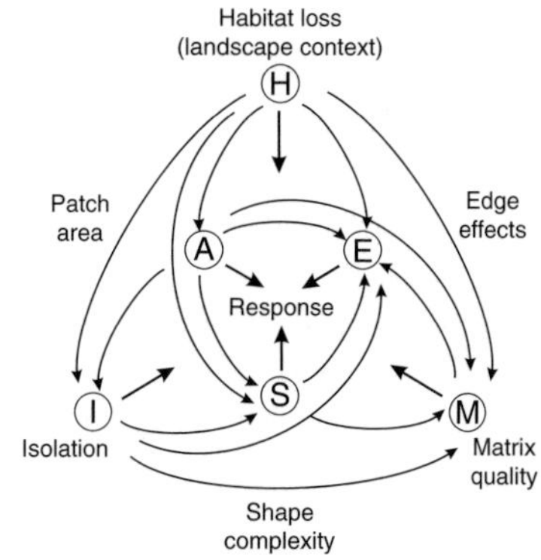

Fig. 4.1. A hierarchical conceptual model of the potential direct and indirect causal paths by which landscape-scale and patch-scale components of habitat fragmentation can affect a measured response variable. The outermost variables of habitat loss (H), matrix quality (M) and isolation (I) are measures of habitat quantity and quality in the landscape surrounding a patch or patches, each of which has patch-level attributes of patch area (A), patch shape complexity (S) and patch edges (E). Arrow heads indicating the direction of causal inference are hypothetical and indicative only, because the direction of causality may vary for a given response variable at a given place or time. (Reprinted, with permission, from Didham *et al.*, 2012b.)

Central to this goal will almost certainly be an even greater focus on the interactions occurring at the intersection of patch-scale and landscape-scale processes, where matrix-context effects have a dominant influence on communities at fragment edges. We discuss the mechanistic drivers of matrix-context effects below, but it is worth pointing out that these have predominantly been studied in comparatively simple landscapes, with low complexity of matrix types and low heterogeneity. In the future, an important goal will be to generalize findings to mosaics of different habitat types, perhaps along the lines of integrating matrix composition and configuration effects as suggested by Fahrig *et al.* (2011).

Mechanistic Drivers of Matrix-context Effects on Ecosystem Dynamics

There are four major mechanisms via which matrix-context effects can influence population or community dynamics: (i) altered movement and dispersal patterns (Ricketts, 2001); (ii) a change in the availability of supplementary or complementary resources (Ries *et al.*, 2004); (iii) altered species interactions (Fagan *et al.*, 1999); and (iv) modification of edge dynamics in remnant patches (Strayer *et al.*, 2003). It is important to recognize that these mechanisms can influence ecological processes across a range of spatial scales. So far, much greater emphasis has been placed on matrix effects on between-patch processes (e.g. Davies *et al.*, 2001) and matrix effects on patch-level dynamics (e.g. Williams *et al.*, 2006; Lindenmayer *et al.*, 2009), whereas relatively few studies have investigated matrix effects on spatially explicit within-patch processes (such as context dependence in edge responses; Campbell *et al.*, 2011).

First, Ricketts (2001) suggested a novel way of conceptualizing patch isolation in terms of the degree to which matrix structure might confer differing levels of 'resistance' to the dispersal patterns of organisms between patches. He demonstrated that the structural complexity of willow versus conifer matrix types had a strong influence on the movement of butterflies dispersing between natural meadow patches, such that matrix type moderated the effect of geographic distance on the 'effective isolation' of habitat patches in fragmented landscapes. Mechanistically, the resistance of matrix habitats can be explained by a number of distinct proximate factors that are rarely discriminated, such as behavioural constraints on dispersing animals or increased mortality rates (Haynes and Cronin, 2006; Laurance *et al.*, 2011), which ultimately determine the rate of movement among patches.

Secondly, the availability, heterogeneity and spatial distribution of resources in the landscape can also play an important role in the response of species to habitat fragmentation. A 'resource', in the broad sense, can be anything from nutrition to specific habitat requirements, and the importance of matrix resources depends on the degree to which organisms might utilize the matrix for purposes other than dispersal. From the perspective of a given organism, Ries and Sisk (2004) suggested that the distribution of resources utilized in landscape mosaics can be either supplementary or complementary among adjacent patch and matrix habitats. The relative distribution of these resources, and their differential utilization by different organisms, might be one of the major factors leading to the seemingly idiosyncratic edge responses often documented in the literature (Murcia, 1995). For example, Ries and Sisk (2004) demonstrated that high availability of supplementary resources between patch and matrix habitats would probably reduce edge contrast effects, but where matrix habitat provides complementary resources (i.e. resources different from those available in the adjacent patch), there would be an expected increase in population abundance around the patch–matrix interface (see Fig. 1 in Ries and Sisk, 2004). By contrast, if the latter example of strong complementarity of resources between patch and matrix is coupled with a high degree of matrix hostility – due to factors such as increased predation rates or climatic extremes – then the matrix can act as a population sink whereby organisms disperse into the matrix to acquire these resources and are then subject to higher mortality rates. Substantially more research needs to be conducted at the interface between behavioural perception of the matrix environment, alteration of top-down predation risk and heterogeneity in bottom-up resource distributions (as well as the 'spillover' of resource subsidies between adjacent habitat types; Rand *et al.*, 2006) before resource-driven effects can be reliably interpreted.

Thirdly, matrix effects influence populations and communities not only via habitat structure and resource availability but also through the alteration of species interactions (Fagan *et al.*, 1999). Altered dispersal patterns due to matrix resistance, increases in matrix-driven mortality and cross-ecosystem subsidies all alter the relative abundances of interacting species across the patch–matrix interface. These effects may act at the population level via density dependence in intraspecific interactions, or at the community level through changes in interspecific species interactions such as predation, competition or mutualism. As a result, species-interaction networks at the landscape scale can be strongly driven by biotic and abiotic factors in the matrix.

Finally, a major determinant of the interaction between patch and matrix habitats is the degree of habitat boundary modification. Strayer *et al.* (2003) suggested that, depending on the characteristics of ecological boundaries, they can be absorptive, reflective or even amplifying. These characteristics include edge 'dimensionality' (such as the width of a buffer zone), patch–matrix contrast (dissimilarity between patch and matrix habitats) and edge 'tortuosity' (the geometrical complexity of edge structure), among others (see Fig. 1 in Strayer *et al.*, 2003). As such, the patch–matrix interface can regulate cross-ecosystem dynamics in landscape mosaics. For example, in landscapes subject to intensive human land uses throughout the matrix, the implementation of regenerating buffer zones surrounding remnant forest patches (to widen edge dimensionality) can reduce negative impacts of threatening matrix processes (Gascon *et al.*, 2000). Moreover, the regulation of production land use (Pawson *et al.*, 2008; Campbell *et al.*, 2011) or the restoration of adjacent matrix habitat can reduce patch–matrix contrast, thus decreasing edge abruptness and potentially reducing the absorptive or reflective properties of the edge.

In order to advance current knowledge of how habitat fragmentation is likely to alter biodiversity and community structure within habitat remnants, a holistic perspective is needed that takes into account these interacting mechanisms at the landscape scale. The shift to a hierarchical landscape perspective on fragmentation effects, combined with clearer mechanistic understanding of the underlying drivers of cross-ecosystem processes, is only now beginning to crystallize into a more structured platform to biodiversity conservation in human-modified landscapes.

Landscape Moderation of Biodiversity Patterns and Processes

In a recent review, Tscharntke *et al.* (2012) synthesized the diverse array of empirical findings about landscape-context effects into six formal ecological hypotheses of landscape moderation of ecosystem dynamics within fragmented landscapes, plus an additional two conservation-focused hypotheses for the way in which landscape-moderated effects might influence management actions. Here, we briefly review the six ecological hypotheses as they scale up from the core concept of cross-habitat spillover to the population-, species- and community-level consequences of landscape-context effects on ecological patterns and processes. Subsequently, we discuss the implications of landscape-moderated effects for biodiversity conservation in human-dominated landscapes (Tscharntke *et al.*, 2012).

The cross-habitat spillover hypothesis

At a fundamental level, ecological processes within forest fragments are inseparably linked to the surrounding matrix. To capture these dynamics, Tscharntke *et al.* (2012) posed the 'cross-habitat spillover hypothesis' (H3 in Tscharntke *et al.*, 2012) as a foundation for exploring how landscape-wide ecological processes are governed by interactions that occur across habitat or ecosystem boundaries. The term 'spillover' emphasizes the importance of flow of resources and organisms across fragment boundaries (Rand *et al.*, 2006). Matrix type and the degree of contrast between fragment and matrix can determine species responses to forest edge environments (Ries *et al.*, 2004; Ewers and Didham, 2006b), as well as the level of cross-habitat movement (Polis *et al.*, 1997; Laurance *et al.*, 2011). Spillover effects occur in both directions across the boundary, with implications for resource availability, species interactions and food web structure in both natural forest fragments and in the adjacent matrix habitat (Ries *et al.*, 2004). Most research on spillover has been conducted from the perspective of how the flow of resources and organisms out of natural systems impacts on population persistence across a network of habitat fragments (Hanski, 1999) or influences production in adjacent agricultural crops (Landis *et al.*, 2000; Tscharntke *et al.*, 2005b). By comparison, relatively little research has focused on spillover from managed to natural systems, or between different natural or different managed systems (Rand *et al.*, 2006; Blitzer *et al.*, 2012), although it could be argued that there is a vast literature on the spread of invasive species throughout modified landscapes that remains to be fully integrated with the habitat fragmentation literature (Didham *et al.*, 2007).

In terms of movement from natural to managed systems, there is wide variation in the propensity for different species to disperse into matrix habitats, utilize matrix resources and survive to reproduce or recolonize other habitat patches. Some species of course are strict old-growth specialists and are rarely observed to occur in the matrix (Gibson *et al.*, 2011). However, this is now recognized to be a somewhat lower proportion of species than previously thought (Edwards *et al.*, 2011). Many forest species exploit the neighbouring matrix to some degree, whether that is just in terms of stochastic dispersal events, supplementary use of matrix resources in addition to those available in forest fragments, or complementary use of distinct resources that are essential for reproduction and are only available in the matrix (e.g. Landis *et al.*, 2000; Knight *et al.*, 2005; Bianchi *et al.*, 2006). In a recent article, Blitzer *et al.* (2012) reviewed the implications of spillover for organisms and their functional roles in five key trophic groups, comprising herbivores, pathogens, pollinators, predators and seed dispersers. For instance, seed dispersal from natural into managed systems is critical for the regeneration of forest in disturbed areas, and Garcia *et al.* (2010) found that the efficiency of seed dispersal by frugivorous birds and bats in human-modified landscapes was closely linked to the proximity of natural habitat remnants. Similarly, Koh (2008) found that bird predation reduced herbivore abundance and herbivory rates in oil palm plantations close to natural forest remnants in South-east Asia, suggesting that protection of the surrounding natural forests is essential for key components of the life histories of many species.

Of the many other examples of spillover effects, perhaps the most widely investigated have been cross-habitat movements of pollinators from natural into managed systems (Garibaldi *et al.*, 2011), with examples available in the literature across many ecosystem types, production systems and biomes (Ricketts *et al.*, 2004; Klein *et al.*, 2007). For example, the spillover of native social bees from tropical forest into coffee agroecosystems has been shown to increase pollination and seed set within 500 m of forest remnants (Klein *et al.*, 2003). However, in such cases, pollinator aggregation in adjacent agricultural crops can also lead to cross-ecosystem competition for pollination services and reduced fitness of native plants in forest remnants, with particular concern over the consequences for rare and threatened native species (Holzschuh *et al.*, 2011). Similarly, Tuell *et al.* (2008) documented flower visitation by managed honeybees on more than half of all monitored native flowers adjacent to the edges of crop fields, with important implications for plant fitness (Lander *et al.*, 2011).

Idiosyncratic effects of spillover on community composition and ecosystem functioning are increasingly being shown across a range of trophic groups, and might largely reflect the novel interactions that result from reciprocal spillover of organisms between managed and natural habitats (Blitzer *et al.*, 2012). For example, Kaiser *et al.* (2008) found that the non-native coffee berry moth, *Prophantis smaragdina* Butler, not only posed a threat to cultivated coffee in Mauritius but also negatively affected the reproductive success of an endemic plant, *Bertiera zaluzania* Comm. ex C.F. Gaertn. in adjacent remnants. Such effects are now apparent across a wide range of taxa, from herbivores (Rand and Louda, 2004) to predators (Lahti, 2001) and pathogens (Wisler and Norris, 2005; Pell *et al.*, 2008; Potts *et al.*, 2010), with a substantial research emphasis now developing on synergistic interactions between landscape change and species invasion into isolated forest remnants (Didham *et al.*, 2007).

The landscape-moderated concentration and dilution hypothesis

The movement of resources or organisms across habitat boundaries can have both transient and long-term effects on populations in forest fragments (Holt and Hochberg, 2001). The 'landscape-moderated concentration and dilution hypothesis' (H4 in Tscharntke *et al.*, 2012) postulates that spatio-temporal changes in land-use and seasonal availability of resources can lead to a temporary increase in population density (crowding) in some forest fragments but a decrease in population density (dilution) in others (Tscharntke *et al.*, 2012). The consequences of transient crowding effects following habitat fragmentation are likely to be most severe for edge-sensitive habitat specialists who depend entirely on the resources found in old-growth forests. Their populations are most vulnerable to sudden elevated competition for limited resources, and cannot be sustained for many generations before numbers collapse. For example, Bierregaard *et al.* (1992) found that bird abundance in isolated fragments increased for approximately 200 days following deforestation of the surrounding habitat, only to drop below pre-fragmentation levels shortly thereafter. Although other examples of transient concentration effects are available in the literature (Tscharntke *et al.*, 2012), most studies rarely have the temporal resolution to measure the time course of population change directly. However, it has also been observed that small isolated remnants frequently have rapid changes in population and community turnover between discrete sampling intervals, suggesting that concentration effects might contribute to 'hyperdynamism' in community composition, and changes in assembly trajectories through time (Laurance *et al.*, 2011).

The dilution effect also describes a shift in the spatial dispersion of a population due to altered movement patterns, but in this case it occurs due to aggregation at the edge, or in the matrix, leading to lower population size within habitat remnants (Tscharntke *et al.*, 2012). Most examples so far have involved the aggregation of species in production crops, such as pollinator aggregation at mass-flowering oilseed rape crops in Germany, leading to reduced pollinator visitation of an endangered plant in adjacent natural grassland (Holzschuh *et al.*, 2011). Relatively few examples of dilution effects are described for forest systems, but one important class of such effects might be 'landscape trap' scenarios in which edges or other modified habitats appear to be preferred by some species, despite increased mortality rates or lower reproductive fitness in these areas (Ries and Fagan, 2003). The general consequence of sink effects operating in landscape traps might be dilution effects within forest remnants. Another potentially unrecognized class of 'functional' dilution effects might be the aggregation of herbivores or pollinators on non-native plants that invade forest habitat, with adverse effects on native plant species. For example, Ghazoul (2004) found that butterfly aggregation on flowers of the non-native invader *Chromolaena odorata* (L.) King & H.E. Robins in logged tropical dry forest in Thailand caused reduced pollinator visitation of *Dipterocarpus obtusifolius* Teijsm. ex Miq. trees.

In general, both concentration and dilution processes are likely to prove more serious for specialist species at higher trophic levels (Tscharntke *et al.*, 2005a) and are likely to have important functional consequences that operate through changes in interaction frequency and trait-dependent shifts in functional roles within ecosystems (Tscharntke *et al.*, 2012).

Landscape-moderated functional trait selection hypothesis

Over longer time periods, fluctuations in the relative abundance of some species compared to others can dramatically alter the trajectory of community assembly (Laurance *et al.*, 2011). For example, in the understorey of rainforest remnants in central Amazonia, plant life-form composition and vegetation structure had transitioned towards a species-poor, disturbed-adapted community in just 9–19 years following fragmentation (Benítez-Malvido and Martínez-Ramos, 2003). The conclusions from a large number of similar studies suggest that the effects of land-use change are highly species specific and depend on landscape type, the spatio-temporal scale of landscape change and variation in species life-history traits (Henle *et al.*, 2004; Ewers and Didham, 2006a; Swift and Hannon, 2010). The 'landscape-moderated functional trait selection hypothesis' (H5 in Tscharntke *et al.*, 2012) describes how habitat fragmentation drives selection for a distinct suite of traits that facilitate survival in human-modified landscapes (Tscharntke *et al.*, 2012).

Among the species at greatest risk from land-use change are large-bodied species, habitat or resource specialists, and those with low reproductive rates (Henle *et al.*, 2004; Cagnolo *et al.*, 2009; Rytwinski and Fahrig, 2011; Canale *et al.*, 2012), whereas matrix-tolerant generalist species tend to have higher survival probabilities in habitat remnants (Laurance *et al.*, 2011). For instance, in a review of the responses of Neotropical vertebrates to habitat fragmentation, Vetter *et al.* (2011) found that species tolerant of open habitats were significantly less affected by habitat fragmentation than others.

The implication of directional trait selection in fragmented systems is that we should be concerned not only with absolute rates of colonization or extinction in forest fragments but also with the potential for non-linear changes in community structure, trophic interactions and ecosystem functioning (Larsen *et al.*, 2005; Lee *et al.*, 2009; Laliberté and Tylianakis; 2010; Williams *et al.*, 2010). In general, the filtering of species that are highly susceptible to fragmentation will cause a shift in species trait distributions (Petchey and Gaston, 2006; Laliberté and Tylianakis, 2011), a loss of response diversity and a reduction in functional redundancy (Laliberté *et al.*, 2010) with important implications for the resilience of ecosystem functioning (Terborgh *et al.*, 2001; Larsen *et al.*, 2005; Tscharntke *et al.*, 2012).

Dominance of the beta diversity hypothesis

Exactly how shifts in population and community processes at the habitat patch level combine to influence biodiversity outcomes at the landscape level across fragmented landscapes is poorly understood (Didham *et al.*, 2012b). Despite overwhelming evidence for the patch-level effects of fragmentation on local biodiversity, the combined effect of local biodiversity losses at many locations might not necessarily be a net loss of biodiversity at the landscape scale in all cases. This is because net biodiversity change at the landscape scale depends as much on patterns of beta diversity (differentiation among habitats) across sites as it does on changes in local alpha diversity (mean species diversity at the habitat level). Tscharntke *et al.* (2012) pose the 'dominance of beta diversity hypothesis' (H2 in Tscharntke *et al.*, 2012) to describe how land-use change might increase levels of beta diversity at larger spatial scales and thereby mitigate the negative impacts of forest fragmentation on alpha diversity at smaller spatial scales. At least part of the explanation for this might be stochasticity, or heterogeneity, in the identities of the species lost at different sites, such that populations might decline locally in distribution even though the species remains 'present' in the landscape until the final population stronghold is extirpated (He and Hubbell, 2011). Recognizing this heterogeneity in population distributions across remnants, and targeting conservation efforts at the resulting patterns of beta diversity in networks of fragments, rather than at alpha diversity within individual fragments, will become increasingly important in fragmented landscapes. For example, in spite of the widely recognized constraints on population persistence in small fragments, Tscharntke *et al.* (2002) found that 29 small grassland patches within a 10 ha area contained more species than did one or two larger patches of the same total area. Similarly, Benedick *et al.* (2006) found substantial beta diversity in butterfly communities among remnant forest patches in a highly fragmented rainforest landscape in northern Borneo, and concluded that small isolated remnants that are often overlooked in conservation planning because of their low alpha diversity can still add substantially to landscape-wide diversity.

Much still remains unclear about the degree of emphasis that should be placed on cumulative measures of landscape-wide biodiversity in conservation planning, because these can be influenced as much by patterns of species gains due to range shifts by generalist native species as by species losses due to the extinction of rare species. Moreover, the level of extinction debt associated with small fragments is almost certainly very high (Kuussaari *et al.*, 2009). In many respects, we believe that beta diversity of patch networks might be best considered from the degree to which heterogeneous patches enhance population persistence of the greatest number of native species.

Landscape-moderated insurance hypothesis

By inference, the dominance of beta diversity hypothesis implies that landscape-level heterogeneity functions as a spatial buffer against local impacts on biodiversity. Whether landscape complexity also buffers the impact of species loss on ecosystem functioning is an important question. To address this, Tscharntke *et al.* (2012) formulated the 'landscape-moderated insurance hypothesis' (H6 in Tscharntke *et al.*, 2012), which posits that landscape complexity provides both spatial and temporal insurance against the impact of land-use change on ecosystem functioning (see also Chapter 7, this volume). Theoretically, high species richness is thought to buffer ecosystem functioning from environmental change if the level of functional redundancy among species within a trophic group is high (Lawton and Brown, 1994; Naeem and Li, 1997; Winfree *et al.*, 2007; Blüthgen and Klein, 2011). Consequently, where local species losses due to fragmentation are very high, and the degree of functional redundancy among remaining taxa is low to moderate, then the resulting impact of fragmentation on ecosystem functioning may be severe. Under these circumstances, Tscharntke *et al.* (2012) emphasized the importance of beta diversity in providing spatial insurance against local declines in ecosystem functioning. Tscharntke *et al.* (2012) considered that the landscape-moderated insurance effect was likely to hold particularly true for dynamic and complex landscapes (Elmqvist *et al.*, 2003) with high levels of beta diversity across multiple scales (Loreau *et al.*, 2003). Many factors will influence the strength of spatial insurance effects in fragmented landscapes, such as the degree of covariance in response-and-effect diversity (i.e. the likelihood that the species responding most sensitively to fragmentation are also those that had the greatest effect on ecosystem functioning) and the degree to which species with similar functional roles can disperse through fragmented landscapes. Nevertheless, the limited evidence available suggests that ecosystem functioning is more resilient to human impacts when local patches are embedded within complex landscapes than within highly simplified landscapes (Tscharntke *et al.*, 2007).

The landscape species pool hypothesis

Finally, the degree to which landscape processes moderate local biodiversity loss and community change, through the mechanisms described above, will be dependent on the size and composition of the regional species pool. This theory suggests that local communities are rarely saturated with species or tightly constrained by local biotic interactions such as competition (Cornell and Lawton, 1992; Gaston, 2000) but are instead open to dispersal and species turnover among localities (Leibold *et al.*, 2004). All other things being equal, the 'landscape species pool hypothesis' (H1 in Tscharntke *et al.*, 2012) predicts that the size of the landscape-wide species pool should moderate local alpha diversity (Tscharntke *et al.*, 2012). In complex species-rich landscapes, there will be a higher probability of stochastic species turnover among localities, and a higher probability of species with suitable traits sorting along altered environmental gradients in fragmented landscapes, thus limiting local biodiversity loss. Conversely, the same degree of habitat loss and alteration of spatial landscape structure might have a much more severe effect on local biodiversity loss when the landscape-wide species pool is small and of limited trait diversity (Bengtsson, 2010; Tscharntke *et al.*, 2012). The relative balance of landscape versus local influences on biodiversity will also depend sensitively on the degree of fragmentation of the remaining habitat. For example, if the landscape is extremely fragmented, there are likely to be few, if any, substantial population 'strongholds' in the remaining patches to provide insurance against further population decline. Below a threshold of habitat fragmentation, negative local effects on communities almost certainly have a flow-on negative influence on landscape-wide biodiversity.

Caveats and Concerns – Are We Underestimating the Severity of Future Biodiversity Losses in Managed Landscapes?

Inevitably, we have only been able to capture a fraction of the potential context dependencies in landscape processes that might influence organisms in fragmented landscapes, and in particular the interactions that almost certainly occur between different processes at different scales. None the less, we believe there is sufficient ecological evidence to suggest that the focal unit of conservation planning should be the wider landscape rather than individual reserved patches. However, the practical allocation of scarce conservation resources within a landscape, framework remains challenging and underexplored (Box 4.2). Should more money be spent on biodiversity on degraded (matrix) land rather than on biodiversity in gazetted nature reserves? Would such an approach be cost-effective and sustainable? The implicit assumption of 'landscape moderation' hypotheses is that the biodiversity values of matrix habitats reflect some form of species resilience to human disturbance and are therefore sustainable through time. A number of lines of evidence suggest that this is unlikely to be the case, and we have serious concerns that the apparent resilience to human disturbance in the matrix belies significant lag effects from past land-use change, as well as significant compounding threats from future land-use intensification and global environmental change.

First, any point assessment of biodiversity in the matrix is but a snapshot of current species distributions, and this fails to take into account spatial and temporal lag effects from past land-use change (Tilman and Downing, 1994; Schrott *et al.*, 2005). Consequently, we are almost certainly underestimating the full effect of historical human impacts on matrix populations over time (Swift and Hannon, 2010). Many species that appear to sustain viable populations in historically modified landscapes remain subject to an extinction debt that is only likely to be paid after several generations (Hanski, 2013). For example, using species richness data for birds in Kenyan forest fragments, Brooks *et al.* (1999) found that fragments retained more species than they were able to support based on species–area predictions, and they concluded that there may be a time lag of up to 50 years before forest fragments lose 50% of the species that are rare and declining to extinction. Similarly, in the more recently deforested areas of the Amazon basin, Wearn *et al.* (2012) found that historical habitat loss has thus far only driven 1% of forest-dependent vertebrate species to extinction, and more than 80% of projected extinctions for rare and declining species are expected to occur over coming decades. These types of extinction lag effects are increasingly well studied in a habitat fragment context but not in a matrix context. Despite this, the same principles almost certainly influence biodiversity in the matrix surrounding isolated remnants, not least because the size and spatial arrangement of patches influences matrix spillover processes as well as extinction debt processes (e.g. Soga and Koike, 2013).

Secondly, spatial dependence in the magnitude of temporal lag effects suggests that local extinctions in degraded matrix habitats may be buffered by landscape-level metapopulation dynamics (Hanski, 2013), but only where sufficient old-growth forest refuges remain in the landscape to act as viable source populations. Exactly what 'sufficient' means in this context is unknown and is in urgent need of empirical evaluation, but it is most likely tied closely to critical thresholds in landscape cover or connectivity that influence population persistence (Fahrig, 2001; Swift and Hannon, 2010). At threshold levels of habitat loss (e.g. >70–90% loss in the landscape; Andrén, 1994), a small additional change in landscape structure can precipitate population collapse in some species. For species that are strongly dependent on old-growth forest habitat, dispersal through the matrix is also likely to decline precipitously near the landscape threshold. For species that can utilize matrix resources to a greater degree, landscape-context effects may buffer populations from old-growth forest loss, but whether this eliminates or simply delays critical threshold effects is unknown. At the extreme, it has even been suggested that negative-feedback processes resulting from critical population thresholds in many interacting species may lead to catastrophic regime shifts due to large-scale habitat loss and associated changes in spatial configuration (Lindenmayer *et al.*, 2008; Pardini *et al.*, 2010).

Thirdly, habitat fragmentation is only one of multiple drivers of global environmental change that interact in complex ways (Didham *et al.*, 2007; Brook *et al.*, 2008) to influence matrix populations. Some of these global change drivers will also have inherently greater impacts in human-modified habitats than in adjacent old-growth remnants, including land-use intensification (Foley *et al.*, 2005), anthropogenic pollution (Körner, 2003; Stevens *et al.*, 2004) and the impact of invasive non-native species (Facon *et al.*, 2006). Non-native species in particular pose a growing threat to natural habitats in modified landscapes, altering ecological interactions and the composition and functioning of invaded ecosystems (Traveset and Richardson, 2006; Pyšek *et al.*, 2012). The apparent resilience of native species to current matrix land use may have no bearing on their long-term persistence at the landscape scale if land-use change is considered in isolation from these other components of global environmental change.

Finally, the potential for conservation gains to be made through biodiversity protection in the matrix will depend on effective policies and governance that limit negative local production practices when local biodiversity impacts become unacceptably high. The problem is that the production of international commodities is increasingly governed by multinational corporations with little understanding or interest in local conservation issues. Consequently, global market drivers of local production practices are spatially decoupled from their negative environmental impacts on biodiversity (Didham *et al.*, 2012a). This decoupling will inevitably lead to increased instability of matrix populations and uncertainty over long-term population persistence into the future.

Conservation Implications

If there is one message that stands out more clearly than any other, it is that the future of forest biodiversity will become increasingly dependent on the ability of species to persist in the face of human impacts, rather than on the ability of conservation management to shield species from those impacts. The wealth of evidence pointing to matrix drivers of landscape-context dependence in patch processes has given rise to renewed hope that sustainable matrix management will mitigate biodiversity losses despite continuing attrition of old-growth forest remnants in the landscape. However, it is worth reiterating that there is a reciprocal dependence between patch and matrix processes. Just as population persistence in a patch network is dependent on matrix processes, so too the persistence of matrix populations is dependent on the amount and configuration of old-growth source populations at a landscape scale. If anything, fragmentation processes will be even more important to consider under a 'biodiversity on degraded lands' paradigm than under a 'wilderness conservation' paradigm (Box 4.2). With intensifying global conflicts between production goals versus biodiversity goals (Godfray *et al.*, 2010), we see three key implications of changing matrix land use for biodiversity conservation in human-dominated landscapes.

First, protected areas are not sufficient in their own right to conserve forest biodiversity (Rodrigues *et al.*, 2004; Bhagwat *et al.*, 2008), particularly as poverty and human population growth are concentrated in many of the areas that are most in need of protection (Wittemyer *et al.*, 2008). We agree that biodiversity conservation inevitably calls for a more integrated landscape approach, as opposed to the segregation of habitats based on land-use types (Daily *et al.*, 2001; Bennett *et al.*, 2006; Vandermeer and Perfecto, 2007; Perfecto and Vandermeer, 2008). This is not to say that large natural forest areas are in any way diminished in their vital importance for the survival of many native species (Turner, 1996; Schulze *et al.*, 2004; FAO, 2006; Barlow *et al.*, 2007; Didham, 2011; Gibson *et al.*, 2011), but they offer only one piece to the biodiversity conservation puzzle. The major challenge will be where to invest limited resources to achieve the greatest conservation benefit.

Tscharntke *et al.* (2012) argue that the greatest conservation gains to be made from matrix management will be in landscapes of intermediate habitat structural complexity rather than in heavily cleared landscapes or near-pristine landscapes (e.g. Tscharntke *et al.*, 2005a; Rundlöf and Smith, 2006; Concepción *et al.*, 2008; Batáry *et al.*, 2010b). For one thing, cleared landscapes with very low (<1%) natural habitat typically do not have the available species pool of organisms to benefit from local conservation management actions (such as habitat restoration or reduction in the intensity of farming practices; Batáry *et al.*, 2010a). At the other extreme, very complex landscapes with a substantial amount of natural habitat area (>20%) typically have high biodiversity and high rates of movement of organisms throughout the landscape, enabling rapid compensation following local human impacts and only a small margin for any biodiversity gains from conservation management actions (Tscharntke *et al.*, 2012). In a recent meta-analysis, Bátary *et al.* (2011) concluded that local agri-environment schemes (such as those supporting a shift from conventional to organic farming practices) predominantly enhanced biodiversity in intermediate-complexity landscapes (1–20% natural habitat area). Moreover, Holzschuh *et al.* (2008) found that multiple conservation management actions at a local scale synergistically enhanced bee diversity across large-scale organic farming landscapes. Whether the conclusions of the 'intermediate landscape-complexity hypothesis' (Tscharntke *et al.*, 2012) are generalizable across farming systems, ecosystem types and geographical regions remains to be investigated.

Secondly, we caution that undue reliance on the conservation gains to be made within matrix habitats is misplaced without explicit consideration of the amount and configuration of old-growth forest remnants within the landscape. The persistence of many species in the matrix is almost certainly subsidized by dispersal ('rescue effects') from old-growth forest habitats, such that the true conservation value of the matrix cannot be determined independently of the fragmentation context. This type of context dependence in the conservation value of matrix habitats is implicit in the intermediate landscape-complexity hypothesis, as well as in the 'landscape-moderated biodiversity versus ecosystem service management hypothesis' (Tscharntke *et al.*, 2012). In the latter hypothesis, Tscharntke *et al.* (2012) postulate that in intensively used agricultural landscapes very few rare or endangered species will survive long term, so the focus should not be on biodiversity conservation per se but rather on the conservation of a wide variety of abundant native species that play important roles in the provision of ecosystem services within agroecosystems (Foster *et al.*, 2011; Kleijn *et al.*, 2011). Pure biodiversity conservation objectives, on the other hand, should focus on landscapes with low-intensity production systems embedded in structurally complex regions with a high proportion of natural habitat remaining (Tscharntke *et al.*, 2012). The relative balance between these differing biodiversity versus ecosystem service outcomes in different landscapes lies at the heart of the land-sparing to land-sharing continuum of land-use trade-offs (Fischer *et al.*, 2008; see also Chapter 7, this volume).

Thirdly, there can be no one-size-fits-all solution to negotiating the trade-offs between biodiversity conservation and commodity production in forested landscapes. Different regions of the world have very different evolutionary histories, different trajectories of land-use change and very different potential for landholders to adopt more sustainable practices that will make a genuine difference to biodiversity (Lindenmayer *et al.*, 2012). It is, therefore, not possible to be broadly prescriptive about which factors will play the most influential role in determining landscape-level solutions to biodiversity loss in managed landscapes. Indeed, many of the landscapes that have the highest levels of biodiversity facing the greatest threat from deforestation are also the regions where our current state of knowledge on existing processes is most limited. The added complexity of future synergies with other components of global environmental change, in all its ecological and political manifestations, only makes our ability to find appropriate solutions to land-use change all the more challenging.

In conclusion, forest fragmentation will remain the dominant threat to global biodiversity into the foreseeable future. Ironically, processes operating in the land-use matrix surrounding forest remnants drive many of the negative impacts of land-use change on remnant forest but at the same time play a vital role in long-term biodiversity conservation in fragmented landscapes. Accordingly, biodiversity conservation must refocus more on the interaction between patch and landscape processes than on patch processes per se. Taking advantage of matrix-context effects will demand more effective discrimination of the relative importance of total habitat loss, declining matrix habitat quality and altered spatial structuring of suitable habitat as mechanistic drivers of biodiversity loss in fragmented forest landscapes.

References

Andrén, H. (1994) Effects of habitat fragmentation on birds and mammals in landscapes with different proportions of suitable habitat: a review. *Oikos* 71, 355–366.

Baker, S.C. and Barmuta, L.A. (2006) Evaluating spatial autocorrelation and depletion in pitfall-trap studies of environmental gradients. *Journal of Insect Conservation* 10, 269–276.

Banks-Leite, C., Ewers, R.M. and Metzger, J.-P. (2010) Edge effects as the principal cause of area effects on birds in fragmented secondary forest. *Oikos* 119, 918–926.

Barlow, J., Gardner, T.A., Araujo, I.S., Ávila Pires, T.C., Bonaldo, A.B., Costa, J.E., Esposito, M.C., Ferreira, L.V., Hawes, J., Hernandez, M.I.M., Hoogmoed, M.S., Leite, R.N., *et al.* (2007) Quantifying the biodiversity value of tropical primary, secondary, and plantation forest. *Proceedings of the National Academy of Sciences USA* 104, 18555–18560.

Batáry, P., Báldi, A., Sárospataki, M., Kohler, F., Verhulst, J., Knop, E., Herzog, F. and Kleijn, D. (2010a) Effect of conservation management on bees and insect-pollinated grassland plant communities in three European countries. *Agriculture, Ecosystems and Environment* 136, 35–39.

Batáry, P., Matthiesen, T. and Tscharntke, T. (2010b) Landscape-moderated importance of hedges in conserving farmland bird diversity of organic vs. conventional croplands and grasslands. *Biological Conservation* 143, 2020–2027.

Bátary, P., Fischer, J., Báldi, A., Crist, T.O. and Tscharntke, T. (2011) Does habitat heterogeneity increase farmland biodiversity? *Frontiers in Ecology and the Environment* 9, 152–153.

Beger, M., Grantham, H.S., Pressey, R.L., Wilson, K., Peterson, E., Dorfman, D., Mumby, P.L., Lourival, R., Brumbaugh, D.R. and Possingham, H.P. (2010) Conservation planning for connectivity across marine, freshwater, and terrestrial realms. *Biological Conservation* 143, 365–575.

Bender, D. and Fahrig, L. (2005) Matrix structure obscures the relationship between interpatch movement and patch size and isolation. *Ecology* 86, 1023–1033.

Benedick, S., Hill, J.K., Mustaffa, N., Chey, V.K., Maryati, M., Searle, J.B., Schilthuizen, M. and Hamer, K.C. (2006) Impacts of rain forest fragmentation on butterflies in northern Borneo: species richness, turnover and the value of small fragments. *Journal of Applied Ecology* 43, 967–977.

Bengtsson, J. (2010) Applied (meta)community ecology: diversity and ecosystem services at the intersection of local and regional processes. In: Verhoef, H.A. and Morin, P.J. (eds) *Community Ecology*. Oxford University Press, Oxford, UK, pp. 115–130.

Benítez-Malvido, J. and Martínez-Ramos, M. (2003) Impact of forest fragmentation on understory plant species richness in Amazonia. *Conservation Biology* 17, 389–400.

Bennett, A.F., Radford, J.Q. and Haslem, A. (2006) Properties of land mosaics: implications for nature conservation in agricultural environments. *Biological Conservation* 133, 250–264.

Bhagwat, S.A., Willis, K.J., Birks, H.J.B. and Whittaker, R.J. (2008) Agroforestry: a refuge for tropical biodiversity? *Trends in Ecology and Evolution* 23, 261–267.

Bianchi, F.J.J.A., Booij, C.J.H. and Tscharntke, T. (2006) Sustainable pest regulation in agricultural landscapes: a review on landscape composition, biodiversity and natural pest control. *Proceedings of The Royal Society B: Biological Sciences* 273, 1715–1727.

Bierregaard, R.O., Lovejoy, T.E., Kapos, V., dos Santos, A.A. and Hutchings, R.W. (1992) The biological dynamics of tropical rainforest fragments. *Bioscience* 42, 859–866.

Blakely, T.J. and Didham, R.K. (2010) Disentangling the mechanistic drivers of ecosystem-size effects on species diversity. *Journal of Animal Ecology* 79, 1204–1214.

Blitzer, E.J., Dormann, C.F., Holzschuh, A., Klein, A.-M., Rand, T.A. and Tscharntke, T. (2012) Spillover of functionally important organisms between managed and natural habitats. *Agriculture, Ecosystems and Environment* 146, 34–43.

Blüthgen, N. and Klein, A.-M. (2011) Functional complementarity and specialisation: the role of biodiversity in plant–pollinator interactions. *Basic and Applied Ecology* 12, 282–291.

Brady, M.J., McAlpine, C.A., Miller, C.J., Possingham, H.P. and Baxter, G.S. (2009) Habitat attributes of landscape mosaics along a gradient of matrix development intensity: matrix management matters. *Landscape Ecology* 24, 879–891.

Brook, B.W., Sodhi, N.S. and Bradshaw, C.J.A. (2008) Synergies among extinction drivers under global change. *Trends in Ecology and Evolution* 23, 453–460.

Brooks, T.M., Pimm, S.L. and Oyugi, J.O. (1999) Time lag between deforestation and bird extinction in tropical forest fragments. *Conservation Biology* 13, 1140–1150.

Bryant, D., Nielsen, D. and Tangley, L. (1997) *The Last Frontier Forests*. World Resources Institute, Washington, DC.

Cagnolo, L., Valladares, G., Salvo, A., Cabido, M. and Zak, M. (2009) Habitat fragmentation and species loss across three interacting trophic levels: effects of life-history and food-web traits. *Conservation Biology* 23, 1167–1175.

Campbell, R.E., Harding, J.S., Ewers, R.M., Thorpe, S. and Didham, R.K. (2011) Production land use alters edge response functions in remnant forest invertebrate communities. *Ecological Applications* 21, 3147–3161.

Canale, G.R., Peres, C.A., Guidorizzi, C.E., Gatto, C.A.F. and Kierulff, M.C.M. (2012) Pervasive defaunation of forest remnants in a tropical biodiversity hotspot. *PLoS ONE* 7, e41671.

Chardon, J.P., Adriaensen, F. and Matthysen, E. (2003) Incorporating landscape elements into a connectivity measure: a case study for the Speckled wood butterfly (*Pararge aegeria* L.). *Landscape Ecology* 18, 561–573.

Chazdon, R.L., Harvey, C.A., Komar, O., Griffith, D.M., Ferguson, B.G., Martínez-Ramos, M., Morales, H., Nigh, R., Soto-Pinto, L., van Breugel, M. and Philpott, S.M. (2009) Beyond reserves: a research agenda for conserving biodiversity in human-modified tropical landscapes. *Biotropica* 41, 142–153.

Collinge, S. (2009) *Ecology of Fragmented Landscapes*. John Hopkins University Press, Baltimore, Maryland.

Collins, C.D., Holt, R.D. and Foster, B.L. (2009) Patch size effects on plant species decline in an experimentally fragmented landscape. *Ecology* 90, 2577–2588.

Concepción, E.D., Díaz, M. and Baquero, R.A. (2008) Effects of landscape complexity on the ecological effectiveness of agri-environment schemes. *Landscape Ecology* 23, 135–148.

Cornell, H.V. and Lawton, J.H. (1992) Species interactions, local and regional processes, and limits to the richness of ecological communities – a theoretical perspective. *Journal of Animal Ecology* 61, 1–12.

Daily, G.C., Ehrlich, P.R. and Sánchez-Azofeifa, G.A. (2001) Countryside biogeography: use of human-dominated habitats by the avifauna of Southern Costa Rica. *Ecological Applications* 11, 1–13.

Davies, K.F., Melbourne, B.A. and Margules, C.R. (2001) Effects of within- and between-patch processes on community dynamics in a fragmentation experiment. *Ecology* 82, 1830–1846.

Didham, R.K. (2010) Ecological consequences of habitat fragmentation. In: *Encyclopedia of Life Sciences*. John Wiley and Sons, Chichester, UK, pp. 1–11.

Didham, R.K. (2011) Life after logging: strategic withdrawal from the Garden of Eden or tactical error for wilderness conservation? *Biotropica* 43, 393–395.

Didham, R.K., Tylianakis, J.M., Gemmell, N.J., Rand, T.A. and Ewers, R.M. (2007) Interactive effects of habitat modification and species invasion on native species decline. *Trends in Ecology and Evolution* 22, 489–496.

Didham, R.K., Denmead, L.H. and Deakin, E.L. (2012a) Riches to rags: the ecological consequences of land-use intensification in New Zealand. In: Lindenmayer, D.B., Cunningham, S.A. and Young, A.D. (eds) *Land Use Intensification: Effects on Agriculture, Biodiversity and Ecological Processes*. CSIRO Publishing, Canberra, pp. 73–83.

Didham, R.K., Kapos, V. and Ewers, R.M. (2012b) Rethinking the conceptual foundations of habitat fragmentation research. *Oikos* 121, 161–170.

Edwards, D.P., Larsen, T.H., Docherty, T.D.S., Ansell, F.A., Hsu, W.W., Derhé, M.A., Hamer, K.C. and Wilcove, D.S. (2011) Degraded lands worth protecting: the biological importance of Southeast Asia's repeatedly logged forests. *Proceedings of The Royal Society B: Biological Sciences* 278, 82–90.

Ellis, E.C., Goldewijk, K.K., Siebert, S., Lightman, D. and Ramankutty, N. (2010) Anthropogenic transformation of the biomes, 1700 to 2000. *Global Ecology and Biogeography* 19, 589–606.

Elmqvist, T., Folke, C., Nyström, M., Peterson, G., Bengtsson, J., Walker, B. and Norberg, J. (2003) Response diversity, ecosystem change, and resilience. *Frontiers in Ecology and the Environment* 1, 488–494.

Ewers, R.M. and Didham, R.K. (2006a) Confounding factors in the detection of species responses to habitat fragmentation. *Biological Reviews* 81, 117–142.

Ewers, R.M. and Didham, R.K. (2006b) Continuous response functions for quantifying the strength of edge effects. *Journal of Applied Ecology* 43, 527–536.

Ewers, R.M., Kliskey, A.D., Walker, S., Rutledge, D., Harding, J.S. and Didham, R.K. (2006) Past and future trajectories of forest loss in New Zealand. *Biological Conservation* 133, 312–325.

Ewers, R.M., Thorpe, S. and Didham, R.K. (2007) Synergistic interactions between edge and area effects in a heavily fragmented landscape. *Ecology* 88, 96–106.

Facon, B., Genton, B.J., Shykoff, J., Jarne, P., Estoup, A. and David, P. (2006) A general eco-evolutionary framework for understanding bioinvasions. *Trends in Ecology and Evolution* 21, 130–135.

Fagan, W.F., Cantrell, R.S. and Cosner, C. (1999) How habitat edges change species interactions. *American Naturalist* 153, 165–182.

Fahrig, L. (1997) Relative effects of habitat loss and fragmentation on population extinction. *Journal of Wildlife Management* 61, 603–610.

Fahrig, L. (2001) How much habitat is enough? *Biological Conservation* 100, 65–74.

Fahrig, L. (2003) Effects of habitat fragmentation on biodiversity. *Annual Review of Ecology, Evolution, and Systematics* 34, 487–515.

Fahrig, L., Baudry, J., Brotons, L., Burel, F.G., Crist, T.O., Fuller, R.J., Sirami, C., Siriwardena, G.M. and Martin, J.L. (2011) Functional landscape heterogeneity and animal biodiversity in agricultural landscapes. *Ecology Letters* 14, 101–112.

FAO (2006) *Global Forest Resources Assessment 2005: Progress towards sustainable forest management*. FAO Forestry Paper 147. Food and Agriculture Organization of the United Nations, Rome.

FAO (2010) *Global Forest Resources Assessment*. Main report. FAO Forestry Paper 163. Food and Agriculture Organization of the United Nations, Rome.

Fischer, J. and Lindenmayer, D.B. (2006) Beyond fragmentation: the continuum model for fauna research and conservation in human-modified landscapes. *Oikos* 112, 473–480.

Fischer, J., Brosi, B., Daily, G.C., Ehrlich, P.R., Goldman, R., Goldstein, J., Lindenmayer, D.B., Manning, A.D., Mooney, H.A., Pejchar, L., Ranganathan, J. and Tallis, H. (2008) Should agricultural policies encourage land sparing or wildlife-friendly farming? *Frontiers in Ecology and the Environment* 6, 380–385.

Foley, J.A., DeFries, R., Asner, G.P., Barford, C., Bonan, G., Carpenter, S.R., Chapin, F.S., Coe, M.T., Daily, G.C., Gibbs, H.K., Helkowski, J.H., Holloway, T. *et al.* (2005) Global consequences of land use. *Science* 309, 570–574.

Foster, W.A., Snaddon, J.L., Turner, E.C., Fayle, T.M., Cockerill, T.D., Ellwood, M.D.F., Broad, G.R., Chung, A.Y.C., Eggleton, P., Khen, C.V. and Yusah, K.M. (2011) Establishing the evidence base for maintaining biodiversity and ecosystem function in the oil palm landscapes of South East Asia. *Philosophical Transactions of The Royal Society B: Biological Sciences* 366, 3277–3291.

Fuller, R.A., McDonald-Madden, E., Wilson, K.A., Carwardine, J., Grantham, H.S., Watson, J.E.M., Klein, C.J., Green, D.C. and Possingham, H.P. (2010) Replacing underperforming protected areas achieves better conservation outcomes. *Nature* 466, 365–367.

Garcia, D., Zamora, R. and Amico, G.C. (2010) Birds as suppliers of seed dispersal in temperate ecosystems: conservation guidelines from real-world landscapes. *Conservation Biology* 24, 1070–1079.

Gardner, T.A., Barlow, J., Chazdon, R., Ewers, R.M., Harvey, C.A., Peres, C.A. and Sodhi, N.S. (2009) Prospects for tropical forest biodiversity in a human-modified world. *Ecology Letters* 12, 561–582.

Garibaldi, L.A., Steffan-Dewenter, I., Kremen, C., Morales, J.M., Bommarco, R., Cunningham, S.A., Carvalheiro, L.G., Chacoff, N.P., Dudenhöffer, J.H., Greenleaf, S.S., Holzschuh, A., Isaacs, R. *et al.* (2011) Stability of pollination services decreases with isolation from natural areas despite honey bee visits. *Ecology Letters* 14, 1062–1072.

Gascon, C., Williamson, G.B. and da Fonseca, G.A.B. (2000) Receding forest edges and vanishing reserves. *Science* 288, 1356–1358.

Gaston, K.J. (2000) Global patterns in biodiversity. *Nature* 405, 220–227.

Ghazoul, J. (2004) Alien abduction: disruption of native plant-pollinator interactions by invasive species. *Biotropica* 36, 156–164.

Gibson, L., Lee, T.M., Koh, L.P., Brook, B.W., Gardner, T.A., Barlow, J., Peres, C.A., Bradshaw, C.J.A., Laurance, W.F., Lovejoy, T.E. and Sodhi, N.S. (2011) Primary forests are irreplaceable for sustaining tropical biodiversity. *Nature* 478, 378–381.

Godfray, H.C.J., Crute, I.R., Haddad, L., Lawrence, D., Muir, J.F., Nisbett, N., Pretty, J., Robinson, S., Toulmin, C. and Whiteley, R. (2010) The future of the global food system. *Philosophical Transactions of The Royal Society B: Biological Sciences* 365, 2769–2777.

Hanski, I. (1999) Habitat connectivity, habitat continuity, and metapopulations in dynamic landscapes. *Oikos* 87, 209–219.

Hanski, I. (2013) Extinction debt at different spatial scales. *Animal Conservation* 16, 12–13.

Harvey, C.A., Komar, O., Chazdon, R., Ferguson, B.G., Finegan, B., Griffith, D.M., Martínez-Ramos, M., Morales, H., Nigh, R., Soto-Pinto, L., van Breugel, M. and Wishnie, M. (2008) Integrating agricultural landscapes with biodiversity conservation in the Mesoamerican hotspot. *Conservation Biology* 22, 8–15.

Haynes, J.K. and Cronin, T.J. (2006) Interpatch movement and edge effects: the role of behavioral responses to the landscape matrix. *Oikos* 113, 43–54.

He, F.L. and Hubbell, S.P. (2011) Species–area relationships always overestimate extinction rates from habitat loss. *Nature* 473, 368–371.

Henle, K., Davies, K.F., Kleyer, M., Margules, C. and Settele, J. (2004) Predictors of species sensitivity to fragmentation. *Biodiversity and Conservation* 13, 207–251.

Hobbs, R.J., Arico, S., Aronson, J., Baron, J.S., Bridgewater, P., Cramer, V.A., Epstein, P.R., Ewel, J.J., Klink, C.A., Lugo, A.E., Norton, D., Ojima, D. *et al.* (2006) Novel ecosystems: theoretical and management aspects of the new ecological world order. *Global Ecology and Biogeography* 15, 1–7.

Holt, R.D. and Hochberg, M.E. (2001) Indirect interactions, community modules and biological control: a theoretical perspective. In: Waijnberg, E., Scott, J.K. and Quimby, P.C. (eds) *Evaluation of Indirect Ecological Effects of Biological Control*. CAB International, Wallingford, UK, pp. 13–37.

Holzschuh, A., Steffan-Dewenter, I. and Tscharntke, T. (2008) Agricultural landscapes with organic crops support higher pollinator diversity. *Oikos* 117, 354–361.

Holzschuh, A., Dormann, C.F., Tscharntke, T. and Steffan-Dewenter, I. (2011) Expansion of mass-flowering crops leads to transient pollinator dilution and reduced wild plant pollination. *Proceedings of The Royal Society B: Biological Sciences* 278, 3444–3451.

Janzen, D.H. (1983) No park is an island – increase in interference from outside as park size decreases. *Oikos* 41, 402–410.

Kaiser, C.N., Hansen, D.M. and Müller, C.B. (2008) Exotic pest insects: another perspective on coffee and conservation. *Oryx* 42, 143–146.

Kareiva, P., Watts, S., McDonald, R. and Boucher, T. (2007) Domesticated nature: shaping landscapes and ecosystems for human welfare. *Science* 316, 1866–1869.

Kleijn, D., Rundlöf, M., Scheper, J., Smith, H.G. and Tscharntke, T. (2011) Does conservation on farmland contribute to halting the biodiversity decline? *Trends in Ecology and Evolution* 26, 474–481.

Klein, A.M., Steffan-Dewenter, I. and Tscharntke, T. (2003) Pollination of *Coffea canephora* in relation to local and regional agroforestry management. *Journal of Applied Ecology* 40, 837–845.

Klein, A.M., Vaissiére, B.E., Cane, J.H., Steffan-Dewenter, I., Cunningham, S.A., Kremen, C. and Tscharntke, T. (2007) Importance of pollinators in changing landscapes for world crops. *Proceedings of The Royal Society B: Biological Sciences* 274, 303–313.

Knight, T.M., McCoy, M.W., Chase, J.M., McCoy, K.A. and Holt, R.D. (2005) Trophic cascades across ecosystems. *Nature* 437, 880–883.

Koh, L.P. (2008) Can oil palm plantations be made more hospitable for forest butterflies and birds? *Journal of Applied Ecology* 45, 1002–1009.

Körner, C. (2003) Ecological impacts of atmospheric CO_2 enrichment on terrestrial ecosystems. *Philosophical Transactions of the Royal Society of London A: Mathematical, Physical and Engineering Sciences* 361, 2023–2041.

Krauss, J., Steffan-Dewenter, I. and Tscharntke, T. (2003) Local species immigration, extinction, and turnover of butterflies in relation to habitat area and habitat isolation. *Oecologia* 137, 591–602.

Kupfer, J.A., Malanson, G.P. and Franklin, S.B. (2006) Not seeing the ocean for the islands: the mediating influence of matrix-based processes on forest fragmentation effects. *Global Ecology and Biogeography* 15, 8–20.

Kuussaari, M., Bommarco, R., Heikkinen, R.K., Helm, A., Krauss, J., Lindborg, R., Ökinger, E., Pärtel, M., Pino, J., Rodá, F., Stefanescu, C., Teder, T. *et al.* (2009) Extinction debt: a challenge for biodiversity conservation. *Trends in Ecology and Evolution* 24, 564–571.

Lahti, D. (2001) The "edge effect on nest predation" hypothesis after twenty years. *Biological Conservation* 99, 365–374.

Laliberté, E. and Tylianakis, J.M. (2010) Deforestation homogenizes tropical parasitoid–host networks. *Ecology* 91, 1740–1747.

Laliberté, E. and Tylianakis, J.M. (2011) Cascading effects of long-term land-use changes on plant traits and ecosystem functioning. *Ecology* 93, 145–155.

Laliberté, E., Wells, J.A., DeClerck, F., Metcalfe, D.J., Catterall, C.P., Queiroz, C., Aubin, I., Bonser, S.P., Ding, Y., Fraterrigo, J.M., McNamara, S., Morgan, J.W. *et al.* (2010) Land-use intensification reduces functional redundancy and response diversity in plant communities. *Ecology Letters* 13, 76–86.

Lander, T., Bebber, D., Choy, C.L., Harris, S. and Boshier, D. (2011) The Circe principle explains how resource-rich land can waylay pollinators in fragmented landscapes. *Current Biology* 21, 1302–1307.

Landis, D.A., Wratten, S.D. and Gurr, G.M. (2000) Habitat management to conserve natural enemies of arthropod pests in agriculture. *Annual Review of Entomology* 45, 175–201.

Larsen, T.H., Williams, N.M. and Kremen, C. (2005) Extinction order and altered community structure rapidly disrupt ecosystem functioning. *Ecology Letters* 8, 538–547.

Laurance, W.F. (2007) Forest destruction in tropical Asia. *Current Science* 93, 1544–1550.

Laurance, W.F. (2008) Theory meets reality: how habitat fragmentation research has transcended island biogeographic theory. *Biological Conservation* 141, 1731–1744.

Laurance, W.F., Camargo, J.L.C., Luizão, R.C.C., Laurance, S.G., Pimm, S.L., Bruna, E.M., Stouffer, P.C., Williamson, G.B., Benítez-Malvido, J., Vasconcelos, H.L., Van Houtan, K.S., Zartman, C.E. *et al.* (2011) The fate of Amazonian forest fragments: a 32-year investigation. *Biological Conservation* 144, 56–67.

Lawton, J.H. and Brown, V.K. (1994) Redundancy in ecosystems. In: Schulze, E.-D. and Mooney, H.A. (eds) *Biodiversity and Ecosystem Function*. Springer, Berlin, pp. 255–270.

Lee, J.S.H., Lee, I.Q.W., Lim, S.L.-H., Huijbregts, J. and Sodhi, N.S. (2009) Changes in dung beetle communities along a gradient of tropical forest disturbance in South-East Asia. *Journal of Tropical Ecology* 25, 677–680.

Leibold, M.A., Holyoak, M., Mouquet, N., Amarasekare, P., Chase, J.M., Hoopes, M.F., Holt, R.D., Shurin, J.B., Law, R., Tilman, D., Loreau, M. and Gonzalez, A. (2004) The metacommunity concept: a framework for multi-scale community ecology. *Ecology Letters* 7, 601–613.

Lindenmayer, D.B. and Fischer, J. (2006) *Habitat Fragmentation and Landscape Change: an Ecological and Conservation Synthesis*. CSIRO Publishing, Collingwood, Victoria, Australia.

Lindenmayer, D., Hobbs, R.J., Montague-Drake, R., Alexandra, J., Bennett, A., Burgman, M., Cale, P., Calhoun, A., Cramer, V., Cullen, P., Driscoll, D., Fahrig, L. *et al.* (2008) A checklist for ecological management of landscapes for conservation. *Ecology Letters* 11, 78–91.

Lindenmayer, D.B., Wood, J.T., Cunningham, R.B., Crane, M., Macgregor, C., Michael, D. and Montague-Drake, R. (2009) Experimental evidence of the effects of a changed matrix on conserving biodiversity within patches of native forest in an industrial plantation landscape. *Landscape Ecology* 24, 1091–1103.

Lindenmayer, D.B., Cunningham, S.A. and Young, A.D. (2012) *Land Use Intensification: Effects on Agriculture, Biodiversity and Ecological Processes*. CSIRO Publishing, Collingwood, Victoria, Australia.

Loreau, M., Mouquet, N. and Gonzalez, A. (2003) Biodiversity as spatial insurance in heterogeneous landscapes. *Proceedings of the National Academy of Sciences USA* 100, 12765–12770.

MacArthur, R.H. and Wilson, E.O. (1967) *The Theory of Island Biogeography*. Princeton University Press, Princeton, New Jersey.

McCauley, D.J. (2006) Selling out on nature. *Nature* 443, 27–28.

Mittermeier, R.A., Mittermeier, C.G., Brooks, T.M., Pilgrim, J.D., Konstant, W.R., da Fonseca, G.A.B. and Kormos, C. (2003) Wilderness and biodiversity conservation. *Proceedings of the National Academy of Sciences USA* 100, 10309–10313.

Murcia, C. (1995) Edge effects in fragmented forests: implications for conservation. *Trends in Ecology and Evolution* 10, 58–62.

Naeem, S. and Li, S. (1997) Biodiversity enhances ecosystem reliability. *Nature* 390, 507–509.

Nicholson, E., Mace, G.M., Armsworth, P.R., Atkinson, G., Buckle, S., Clements, T., Ewers, R.M., Fa, J.E., Gardner, T.A., Gibbons, J., Grenyer, R., Metcalfe, R. *et al.* (2009) Priority research areas for ecosystem services in a changing world. *Journal of Applied Ecology* 46, 1139–1144.

Pardini, R., de Arruda Bueno, A., Gardner, T.A., Prado, P.I. and Metzger, J.P. (2010) Beyond the fragmentation threshold hypothesis: regime shifts in biodiversity across fragmented landscapes. *PLoS ONE* 5, e13666.

Pawson, S.M., Brockerhoff, E.G., Meenken, E.D. and Didham, R.K. (2008) Non-native plantation forests as alternative habitat for native forest beetles in a heavily modified landscape. *Biodiversity and Conservation* 17, 1127–1148.

Pell, J.K., Baverstock, J., Roy, H.E., Ware, R.L. and Majerus, M.E. (2008) Intraguild predation involving *Harmonia axyridis*: a review of current knowledge and future perspectives. *Biological Control* 53, 147–168.

Perfecto, I. and Vandermeer, J. (2008) Biodiversity conservation in tropical agroecosystems: a new conservation paradigm. *Year in Ecology and Conservation Biology 2008* 1134, 173–200.

Petchey, O.L. and Gaston, K.J. (2006) Functional diversity: back to basics and looking forward. *Ecology Letters* 9, 741–758.

Polis, G.A., Anderson, W.B. and Holt, R.D. (1997) Toward an integration of landscape and food web ecology: the dynamics of spatially subsidized food webs. *Annual Review of Ecology and Systematics* 28, 289–316.

Potts, S.G., Biesmeijer, J.C., Kremen, C., Neumann, P., Schweiger, O. and Kunin, W.E. (2010) Global pollinator declines: trends, impacts and drivers. *Trends in Ecology and Evolution* 25, 345–353.

Prugh, L.R., Hodges, K.E., Sinclair, A.R.E. and Brashares, J.S. (2008) Effect of habitat area and isolation on fragmented animal populations. *Proceedings of the National Academy of Sciences USA* 105, 20770–20775.

Pyšek, P., Jarošik, V., Hulme, P.E., Pergl, J., Hejda, M., Schaffner, U. and Vila, M. (2012) A global assessment of invasive plant impacts on resident species, communities and ecosystems: the interaction of impact measures, invading species' traits and environment. *Global Change Biology* 18, 1725–1737.

Rand, T.A. and Louda, S.M. (2004) Exotic weed invasion increases the susceptibility of native plants attack by a biocontrol herbivore. *Ecology* 85, 1548–1554.

Rand, T.A., Tylianakis, J.M. and Tscharntke, T. (2006) Spillover edge effects: the dispersal of agriculturally subsidized insect natural enemies into adjacent natural habitats. *Ecology Letters* 9, 603–614.

Ricketts, T.H. (2001) The matrix matters: effective isolation in fragmented landscapes. *American Naturalist* 158, 87–99.

Ricketts, T.H., Daily, G.C., Ehrlich, P.R. and Michener, C.D. (2004) Economic value of tropical forest to coffee production. *Proceedings of the National Academy of Sciences USA* 101, 12579–12582.

Ries, L. and Fagan, W.F. (2003) Habitat edges as a potential ecological trap for an insect predator. *Ecological Entomology* 28, 567–572.

Ries, L. and Sisk, T.D. (2004) A predictive model of edge effects. *Ecology* 85, 2917–2926.

Ries, L., Fletcher, R.J., Battin, J. and Sisk, T.D. (2004) Ecological responses to habitat edges: mechanisms, models, and variability explained. *Annual Review of Ecology, Evolution, and Systematics* 35, 491–522.

Rodrigues, A.S.L., Andelman, S.J., Bakarr, M.I., Boitani, L., Brooks, T.M., Cowling, R.M., Fishpool, L.D.C., da Fonseca, G.A.B., Gaston, K.J., Hoffmann, M., Long, J.S., Marquet, P.A. *et al.* (2004) Effectiveness of the global protected area network in representing species diversity. *Nature* 428, 640–643.

Rudel, T.K., Defries, R., Asner, G.P. and Laurance, W.F. (2009) Changing drivers of deforestation and new opportunities for conservation. *Conservation Biology* 23, 1396–1405.

Rundlöf, M. and Smith, H.G. (2006) The effect of organic farming on butterfly diversity depends on landscape context. *Journal of Applied Ecology* 43, 1121–1127.

Rytwinski, T. and Fahrig, L. (2011) Reproductive rate and body size predict road impacts on mammal abundance. *Ecological Applications* 21, 589–600.

Schrott, G.R., With, K.A. and King, A.W. (2005) On the importance of landscape history for assessing extinction risk. *Ecological Applications* 15, 493–506.

Schulze, C.H., Waltert, M., Kessler, P.J.A., Pitopang, R., Veddeler, D., Mühlenberg, M., Gradstein, S.R., Leuschner, C., Steffan-Dewenter, I. and Tscharntke, T. (2004) Biodiversity indicator groups of tropical land-use systems: comparing plants, birds, and insects. *Ecological Applications* 14, 1321–1333.

Seabloom, E.W., Dobson, A.P. and Stoms, D.M. (2002) Extinction rates under nonrandom patterns of habitat loss. *Proceedings of the National Academy of Sciences USA* 99, 11229–11234.

Soga, M. and Koike, S. (2013) Mapping the potential extinction debt of butterflies in a modern city: implications for conservation priorities in urban landscapes. *Animal Conservation* 16, 1–11.

Stevens, C.J., Dise, N.B., Mountford, J.O. and Gowing, D.J. (2004) Impact of nitrogen deposition on the species richness of grasslands. *Science* 303, 1876–1879.

Strayer, D.L., Power, M.E., Fagan, W.F., Pickett, S.T.A. and Belnap, J. (2003) A classification of ecological boundaries. *Bioscience* 53, 723–729.

Struebig, M.J., Kingston, T., Zubaid, A., Mohd-Adnan, A. and Rossiter, S.J. (2008) Conservation value of forest fragments to palaeotropical bats. *Biological Conservation* 141, 2112–2126.

Swift, T.L. and Hannon, S.J. (2010) Critical thresholds associated with habitat loss: a review of the concepts, evidence, and applications. *Biological Reviews* 85, 35–53.

Tabarelli, M. (2010) Tropical biodiversity in human-modified landscapes: what is our trump card? *Biotropica* 42, 553–554.

Terborgh, J., Lopez, L., Nuñez, P., Rao, M., Shahabuddin, G., Orihuela, G., Riveros, M., Ascanio, R., Adler, G.H., Lambert, T.D. and Balbas, L. (2001) Ecological meltdown in predator-free forest fragments. *Science* 294, 1923–1926.

Tilman, D. and Downing, J.A. (1994) Biodiversity and stability in grasslands. *Nature* 367, 363–365.

Traveset, A. and Richardson, D.M. (2006) Biological invasions as disruptors of plant reproductive mutualisms. *Trends in Ecology and Evolution* 21, 208–216.

Tscharntke, T., Steffan-Dewenter, I., Kruess, A. and Thies, C. (2002) Contribution of small habitat fragments to conservation of insect communities of grassland-cropland landscapes. *Ecological Applications* 12, 354–363.

Tscharntke, T., Klein, A.M., Kruess, A., Steffan-Dewenter, I. and Thies, C. (2005a) Landscape perspectives on agricultural intensification and biodiversity – ecosystem service management. *Ecology Letters* 8, 857–874.

Tscharntke, T., Rand, T.A. and Bianchi, F.J.J.A. (2005b) The landscape context of trophic interactions: insect spillover across the crop–noncrop interface. *Annales Zoologici Fennici* 42, 421–432.

Tscharntke, T., Bommarco, R., Clough, Y., Crist, T.O., Kleijn, D., Rand, T.A., Tylianakis, J.M., van Nouhuys, S. and Vidal, S. (2007) Conservation biological control and enemy diversity on a landscape scale. *Biological Control* 43, 294–309.

Tscharntke, T., Tylianakis, J.M., Rand, T.A., Didham, R.K., Fahrig, L., Batáry, P., Bengtsson, J., Clough, Y., Crist, T.O., Dormann, C.F., Ewers, R.M., Fründ, J. *et al.* (2012) Landscape moderation of biodiversity patterns and processes – eight hypotheses. *Biological Reviews* 87, 661–685.

Tuell, J.K., Fiedler, A.K., Landis, D. and Isaacs, R. (2008) Visitation by wild and managed bees (Hymenoptera: Apoidea) to Eastern U.S. native plants for use in conservation programs. *Environmental Entomology* 37, 707–718.

Turner, I.M. (1996) Species loss in fragments of tropical rain forest: a review of the evidence. *Journal of Applied Ecology* 33, 200–209.

Vandermeer, J. and Perfecto, I. (2007) The agricultural matrix and a future paradigm for conservation. *Conservation Biology* 21, 274–277.

Vetter, D., Hansbauer, M.M., Végvári, Z. and Storch, I. (2011) Predictors of forest fragmentation sensitivity in neotropical vertebrates: a quantitative review. *Ecography* 34, 1–8.

Wearn, O.R., Reuman, D.C. and Ewers, R.M. (2012) Extinction debt and windows of conservation opportunity in the Brazilian Amazon. *Science* 337, 228–232.

Wiens, J.A. (2009) Landscape ecology as a foundation for sustainable conservation. *Landscape Ecology* 24, 1053–1065.

Wilcove, D.S., McLellan, C.H. and Dobson, A. (1986) Habitat fragmentation in the temperate zone. In: Soule, M.E. (ed.) *Conservation Biology. The Science of Scarcity and Diversity*. Sinauer Associates, Sunderland, Massachusetts, pp. 237–256.

Williams, N.M., Crone, E.E., Roulston, T.H., Minckley, R.L., Packer, L. and Potts, S.G. (2010) Ecological and life-history traits predict bee species responses to environmental disturbances. *Biological Conservation* 143, 2280–2291.

Williams, N.S.G., Morgan, J.W., McCarthy, M.A. and McDonnell, M.J. (2006) Local extinctions of grassland plants: the landscape matrix is more important than patch attributes. *Ecology* 87, 3000–3006.

Winfree, R., Griswold, T. and Kremen, C. (2007) Effect of human disturbance on bee communities in a forested ecosystem. *Conservation Biology* 21, 213–223.

Wisler, G.C. and Norris, R.F. (2005) Interactions between weeds and cultivated plants as related to management of plant pathogens. *Weed Science* 53, 914–917.

Wittemyer, G., Elsen, P., Bean, W.T., Burton, A.C.O. and Brashares, J.S. (2008) Accelerated human population growth at protected area edges. *Science* 321, 123–126.

5 Forest Fragmentation Genetics: What Can Genetics Tell Us About Forest Fragmentation?

Aline Finger,[1]* Ute Radespiel,[2] Jan C. Habel[3] and Chris J. Kettle[4]

[1]*Royal Botanic Garden Edinburgh, UK;* [2]*University of Veterinary Medicine Hanover, Germany;* [3]*Technische Universität München, Freising, Germany;* [4]*ETH Zurich, Switzerland*

Abstract

Worldwide, forest biodiversity and other ecosystem services are threatened by anthropogenic-induced forest fragmentation at multiple scales and intensities. Loss of habitat is just one aspect of the overall fragmentation process. The consequences of fragmentation depend on how different organisms respond to the intervening matrix and how complex interactions between species and species assemblages are disrupted. These processes are often cryptic or difficult to observe and remain largely unpredictable. Molecular approaches provide a powerful tool for investigating critical processes such as species mating behaviour or dispersal following forest fragmentation. The erosion of genetic diversity through population isolation and restriction of gene flow in fragmented forests can also be accurately estimated. Forest fragmentation has been shown to have negative genetic impacts on some plant and animal species by increasing mating between close relatives (inbreeding), thus reducing their overall genetic diversity or decreasing population viability. Nevertheless, some species do not seem to be negatively affected by forest fragmentation, for example either because they manage to disperse through a wide range of landscape elements or because they are adapted to persist in small and isolated habitats. It can be difficult to detect genetic consequences of fragmentation, particularly when fragmentation has been relatively recent and species have long generation times. In this chapter, we review the current state of knowledge on forest fragmentation genetics with a focus on plant and animal species, highlight the importance of molecular approaches for informing management of forest fragments and discuss priority areas for future research.

Introduction

Long-term threats to global biodiversity result not only from habitat loss per se but also from negative genetic impacts, which may undermine the ability of species to survive (Frankham, 1995; Young *et al.*, 1996; Ross *et al.*, 2002). It is apparent, however, that the effects of forest fragmentation can be highly idiosyncratic and species specific, with some species showing negative ecological and genetic effects and others showing seemingly no effect at all. This chapter focuses on genetic consequences caused by anthropogenic-induced forest fragmentation and provides an overview of our empirical understanding

*a.finger@rbge.ac.uk

of fragmentation for populations of a wide range of forest-dependent taxa. We will illustrate how tree species – as the foundation of forests – might be affected, followed by studies from invertebrate and vertebrate taxa, which may also play an important role as the predominant pollinator or seed vectors in moist tropical forests. We will explore the power of molecular approaches to study: (i) genetic diversity, inbreeding and population bottlenecks: (ii) the importance of a species historical distribution for its sensitivity to fragmentation; (iii) gene flow, genetic differentiation and mating systems; and (iv) methodological limits to detecting genetic consequences of forest fragmentation. The importance of genetics for the long-term conservation of forest biodiversity and priority areas for future studies are also discussed.

Forest fragmentation from a genetic perspective

Fragmentation of continuous forests to isolated patches can greatly influence plant and animal compositions due to multiple factors. Critical factors described by Fahrig (2003) included: (i) a reduction in forest surface area; (ii) an increase in the number of fragments; (iii) a decrease in fragment size; and (iv) an increase in fragment isolation. All of these factors will affect species, their interactions and their genetic constitution. Population isolation, caused by altered landscape elements, has been shown to change population genetic composition, demography and mating patterns, increasing mating events between related individuals (inbreeding), and to reduce genetic diversity and viability of forest tree species (Bacles *et al.*, 2005; Ghazoul, 2005; Lowe *et al.*, 2005), other plant species (Cuartas-Hernandez and Nunez-Farfan, 2006; Vandepitte *et al.*, 2009) and animals (birds: Brown *et al.*, 2004; Woltmann *et al.*, 2012; mammals: Craul *et al.*, 2009; Haag *et al.*, 2010). Whereas the loss of genetic diversity may impede a species' ability to adapt, and thus survive in changing environmental conditions, inbreeding is known to have a more immediate impact on population viability (Ghazoul *et al.*, 1998; Cascante *et al.*, 2002; Sebbenn *et al.*, 2011). Genetic drift and inbreeding can lead to the enrichment or even fixation of recessive deleterious alleles, leading to a further reduction of fitness through 'genetic erosion' (Van Geert *et al.*, 2008). As a consequence, inbreeding depression – a reduced reproductive fitness following inbreeding – poses a long-term threat to populations and species. Finally, populations with low genetic diversity are expected to suffer more seriously from novel pests, pathogens and parasites, and from demographic and environmental stochasticity, than those with high genetic diversity (Namkoong, 1991; Burdon, 2001; Spielman *et al.*, 2004; Theodorou *et al.*, 2009).

Inbreeding depression and low genetic diversity are expected to be more pronounced in small rather than in large populations (Reed, 2005; Leimu *et al.*, 2006). Therefore, forest fragmentation that leads to small and isolated forest patches could be particularly problematic for forest-dependant species. Rare species are predicted to have a lower genetic diversity than common sister taxa (Hamrick and Godt, 1996; Glitzendanner and Soltis, 2000; Cole, 2003), due to an intrinsically small population size. This leads to the hypothesis that forest fragmentation may have particularly negative genetic effects on rare species. In contrast to this idea, a recent meta-analysis of genetic consequences of habitat fragmentation in plants suggested that common species may be as susceptible to the deleterious genetic consequences of fragmentation as rare species (Honnay and Jacquemyn, 2007), although species-specific responses appear to be moderated by life history, with outcrossing species likely to be more susceptible to the deleterious genetic consequences of habitat degradation (Ward *et al.*, 2005; Aguilar *et al.*, 2006). Empirical studies are advancing our knowledge, for example showing that highly self-compatible tree species may be less vulnerable in the short term (Finger *et al.*, 2012), while predominately outcrossing species may be more vulnerable (see references in Ng *et al.*, 2009; Ismail *et al.*, 2012). These studies highlight the importance of identifying species-specific traits that make them particularly vulnerable to genetic effects associated with forest fragmentation to provide a better understanding for species conservation and management.

Different rates and scales of fragmentation: tropical versus temperate forests

Changes in land-use and unsustainable logging practices have resulted in a substantial degree of forest fragmentation, not only in the tropics but in many parts of the world (Geburek and Konrad, 2008) (Fig. 5.1). Wade *et al.* (2003, p. 1) demonstrated that 'Globally, over half of the temperate broadleaf and mixed forest biome and nearly one quarter of the tropical rainforest biome have been fragmented or removed by humans, as opposed to only 4% of the boreal forest. Overall, Europe had the most human-caused fragmentation and South America the least'. Yet other studies show that boreal forests have experienced the greatest area of forest lost in recent decades (Hansen *et al.*, 2010).

Contemporary responses of organisms to fragmentation may be related to their past fragmentation history. One of the important differences between temperate and tropical regions is that, in general, tropical forests have only recently become fragmented (but see Bhagwat, Chapter 2, this volume) and fragmentation is more rapid and acute (Wright and Muller-Landau, 2006). In contrast, large areas of temperate forested regions have experienced a long history of fragmentation, driven by human settlement, agricultural expansion and economic developments over the last two millennia, following the classical forest transition curve. The rates of forest fragmentation across Europe were relatively slow due to technological limitation and much of

Fig. 5.1. View of the forest landscape of Koorg, Western Ghats, India, showing historic and more recent forest fragmentation due to changing agricultural practices. (© Chris J. Kettle.)

the forest clearing being by manual labour. Conversely, tropical regions have seen an explosion of land use change in the last 50 years (see Laurance, Chapter 3, this volume). Moreover, tropical forests have been diversity hotspots for more than 250 million years and both a cradle and museum of diversity (Jablonski *et al.*, 2006). To what extent these old autochthonous tropical forests are more sensitive to fragmentation, compared with comparatively young low diversity temperate forests, is still not clear. Temperate forests are more strongly impacted by changing climates and are naturally in a constant process of colonization and retreat, depending on climatic conditions, such as glacial periods in North America and Europe. This could lead to the assumption that temperate forests and their constituent tree species may be better adapted to changing environmental conditions and there could be less negative genetic impacts from forest fragmentation. But so far, there has been no study looking at this aspect particularly. Recent papers focusing on how tropical forests are likely to respond to future anthropogenic change (Corlett, 2011) or how past climate change influences their vulnerability (Cannon *et al.*, 2009) highlight the limits to our knowledge, especially regarding thermal tolerance of species and assisted migration. Furthermore, forest fragments are important for mitigation of climate change (Corlett, Chapter 6, this volume), which highlights the need to increase our understanding on the long-term viability of many forest fragments.

Many tropical species will have to be able to survive in highly fragmented landscapes. It is pertinent to note that in tropical forests (in contrast to temperate forests) the interactions between different taxa can be crucial to the different genetic consequence of fragmentation; for example, the preponderance of tropical tree species with insect or animal pollination and animal dispersed fruits (zoochory), in contrast to forest tree species in temperate regions many of which are wind pollinated and have passively or wind-dispersed seeds. From a genetic perspective it is likely that mobile species or species with a broad habitat range and the capability of long-distance dispersal will exhibit less negative genetic effects, as these traits will prevent genetic isolation. Nevertheless, the magnitude of the genetic consequences of habitat fragmentation depends on multiple factors including: population history, mating system, ability to purge deleterious alleles following inbreeding and gene dispersal ability. In addition, synergistic effect, such as increased hunting pressure associated with increased fragmentation, may lead to a loss of important seed dispersal agents, irrespective of their ability to disperse across a matrix (McConkey *et al.*, 2012). There is an urgent need for research to determine whether species with certain levels of endemism or habitat ranges are particularly threatened.

Forest Fragmentation Genetics: Implication for Forest Trees and Other Forest Biodiversity

The field of molecular ecology has enabled ecologists to investigate processes that are either very difficult or near impossible to study using basic ecological survey techniques. For example, across all taxa, analysis of population structure, mating system, gene flow, dispersal over time, inbreeding, genetic diversity and estimates of the effective population sizes can all be determined using molecular approaches. This provides valuable information on population and species vulnerability to forest fragmentation. In the following sections we highlight the major advances of using genetic methods and the critical processes that can be altered by forest fragmentation in both plant and animal species.

Genetic diversity and inbreeding

Genetic diversity comprises the differences in gene composition of individuals within a species. Theoretically, high genetic diversity should be found in non-threatened and widespread species (due to their numbers alone), and a low genetic diversity is expected in rare, endangered and endemic species (Soulé, 1976; Karron, 1991; Frankham, 1996; Glitzendanner and Soltis, 2000; Allendorf and Luikart, 2007).

Inbreeding and the loss of genetic diversity are more likely in small isolated populations, and threatened or rare species typically have small or declining populations (Boxes 5.1 and 5.2). Populations may be isolated geographically but also genetically, and in many cases it is difficult to differentiate between these processes without using molecular methods as, for example, geographically isolated populations may still be in genetic exchange and therefore not reproductively isolated.

That a loss of genetic diversity may lead to extinction in the long term has been argued using theoretical studies (Finke and Jetschke, 1999; Fowler and Whitlock, 1999; Lande, 1988; Mills and Smouse, 1994; Tanaka, 2000), laboratory experiments (Frankham, 1995; Bryant *et al.*, 1999; Bijlsma *et al.*, 2000; Reed and Bryant, 2000; Nieminen *et al.*, 2001), field experiments (Newman and Pilson, 1997), empirical studies on metapopulation dynamics in nature (Saccheri *et al.*, 1998) and meta-analyses (Frankham, 1999). Genotypic diversity was also found to be important for ecosystem function and recovery (Reusch *et al.*, 2005) and for maintaining evolutionary potential in populations (Reed and Frankham, 2003). This underlies the importance of genetic diversity as an essential component of biodiversity (in addition to species and ecosystem diversity) requiring protection (McNeely *et al.*, 1990), although this remains an underevaluated component of biodiversity (Laikre *et al.*, 2010).

Importance of a species' historical distribution for its sensitivity to fragmentation

Species natural distributions can largely be explained by their biogeographical histories. While some species are restricted to certain environmental conditions (e.g. soil types, climatic conditions, availability of food sources), other species' distributions may be related to their dispersal history over past generations. Genetic data can be used to decipher a species'

Box 5.1. Forest fragmentation, genetic erosion and population extinction in plants

Empirical research has demonstrated substantial evidence to support the relationship between population size, genetic diversity, inbreeding and reduced fitness in wild plant populations (O'Grady *et al.*, 2004; Lowe *et al.*, 2005; Leimu *et al.*, 2006). Indeed, Honnay and Jacquemyn (2007) found a positive correlation between population size and genetic diversity. Other meta-analyses support this finding, stating that population size is positively correlated with genetic diversity and fitness (Reed, 2005; Leimu *et al.*, 2006). Such declines in reproduction that are disproportionate to the declines in population size (Allee effects: Stephens *et al.*, 1999) can lead to reduced efficiency of pollination and seed dispersal processes in plant species (reviewed by Ghazoul, 2005). Still, there seems to be controversy as to whether rare, widespread, endemic or non-endemic species will be more prone to negative genetic consequences from forest fragmentation. Common species and large populations were found to be as susceptible, or even more susceptible, to the loss of genetic diversity through fragmentation compared with rare species and small populations (Honnay and Jacquemyn, 2007; Angeloni *et al.*, 2011).

There is, however, variability in the genetic responses of species to forest fragmentation. Tree species, for example, have been suggested to be especially resistant to negative genetic effects following forest fragmentation (Hamrick, 2004), although Angeloni *et al.* (2011) demonstrated that tree species express more inbreeding depression following fragmentation than shrubs and herbs. Indeed, a study of the endangered Indian tropical timber tree *Dysoxylum malabaricum* (Meliaceae) demonstrated that negative genetic effects can even occur in species with long-distance dispersal potential. Here, low population density associated with forest fragmentation resulted in an increase of short-distance mating and average kinship among mates (Ismail *et al.*, 2012). Nursery experiments showed that this elevation in mating between related individuals was associated with reduced fitness (Ismail, 2013). Similar negative effects have been detected for the dipterocarp tree *Vateriopsis seychellarum*, endemic to the Seychelles, which has suffered an extreme reduction in numbers due to forest fragmentation. A decline in genetic diversity and an increase in genetic differentiation result from genetic isolation due to low gene flow rates among local populations (Finger *et al.*, 2012).

Box 5.2. Forest fragmentation, genetic erosion and population extinction in animals

An increasing loss of genetic diversity and increasing genetic differentiation in forest fragments of decreasing size has also been demonstrated in several animal models, such as reptiles (Malone *et al.*, 2003), birds (Oliveras de Ita *et al.*, 2012) and mammals (Sommer, 2003; Potter *et al.*, 2012). Forest fragmentation has led to the loss of genetic diversity, even for species that are potentially robust to fragmentation. The Australian common ringtail possum (*Pseudocheirus peregrinus*), for example, had lower heterozygosity and allelic richness in smaller forest patches, despite its known use of the surrounding forest patch matrix (pine plantations) for foraging and nesting (Lancaster *et al.*, 2011). In British Columbia, forest fragmentation has led to reduced genetic diversity in larvae of the tailed frog (*Ascaphus truei*) compared with old-growth forests (Wahbe *et al.*, 2005). Negative effects of inbreeding were also detected on the immunocompetence and survival, i.e. fitness, in a study of semi-freeliving ring-tailed lemurs (*Lemur catta*) (Charpentier *et al.*, 2008). The authors showed that homozygosity negatively influenced the prevalence of a parasitic fly and hampered survival in both sexes, reducing the median survival age (age with 50% probability of survival) of the more homozygous individuals to less than half of that of the heterozygous individuals.

Furthermore, a meta-analysis by Garner *et al.* (2005) compared the expected heterozygosity of populations stemming from 108 mammalian species that were either 'demographically challenged' (defined as having experienced a population decline, a reduction in population range or isolation from conspecifics) or 'demographically healthy' (defined as not having experienced any of these factors). They showed a consistent and significant loss of genetic diversity (i.e. heterozygosity) in populations that had experienced a demographic threat. These populations showed over a 20% decline in heterozygosity compared with 'healthy' populations, independent of body mass. Similar conclusions were drawn in a meta-analysis on bird genetic diversity (heterozygosity) and its relationship to varying population size (Evans and Sheldon, 2008) in a sample of 89 species. The authors found a significant positive correlation between population size and heterozygosity. Population size is one of the first variables that can be expected to decrease directly as a consequence of fragmentation events. The authors of both meta-analyses, however, did not explicitly link their meta-analyses to fragmentation events.

Direct molecular effects of population bottlenecks (survival of only a few individuals) on genetic diversity are provided for example by various studies on primates (e.g. orang-utan, golden-brown mouse lemur, Goossens *et al.*, 2006; Olivieri *et al.*, 2008), other mammals (e.g. Japanese field mouse, Hirota *et al.*, 2004), birds (e.g. black grouse, Caizergues *et al.*, 2003) and amphibians (e.g. European tree frogs, Andersen *et al.*, 2004). Recent Bayesian methods allow us even to date detected past bottlenecks, for example in the case of orang-utan (Goossens *et al.*, 2006), in which a massive population collapse in a series of forest fragments along the Kinabatangan River in eastern Sabah most probably started within the last few decades, which coincides with the period of intensified forest fragmentation in this region.

evolutionary history. The distribution of different genotypes in populations over large geographical scales and the effects of genetic bottlenecks can reflect species' historical dispersal (Avise, 2000; Moritz, 1995). Limited dispersal may be particularly problematic for species when it comes to forest fragmentation. The inability to colonize new habitats presents novel challenges in conservation, leading to management strategies such as assisted migration, the need for transplantations and habitat restorations (Mansourian *et al.*, 2005; Loss *et al.*, 2011). Information on species' ability to disperse in time and space is crucial to evaluate current risks for species and their potential for natural population recovery or recolonization.

Only a few studies have explored the importance of historical distribution patterns for a plant species' vulnerability to the negative consequences of forest fragmentation (Boxes 5.3. and 5.4) (but see Angeloni *et al.*, 2011; Brigham, 2003; Leimu and Mutikainen, 2005). For example, inbreeding depression may be low in populations that have naturally small patchy distributions and have persisted as such for a long time, and where deleterious alleles have been purged. In contrast, a recent reduction in population size may cause stronger inbreeding depression (Lande and Schemske, 1985; Keller and Waller, 2002) in species that are naturally widespread. Thus, species that have survived in naturally small and isolated populations over long time

Box 5.3. Historical species distribution and genetic response to fragmentation in plants

Not all plant species seem to be negatively affected by fragmentation; the differential responses are driven largely by the diverse range of reproductive strategies that plants exhibit. For example, naturally scattered species may be less vulnerable to negative genetic effects in fragmented environments. *Cedrus brevifolia* (Pinaceae), a tree species that is naturally restricted to a small region in Cyprus but has been further reduced due to anthropogenic causes, has retained a high genetic diversity and has low inbreeding and genetic differentiation (Eliades *et al.*, 2011). Similarly, *Lepidorrhachis mooreana* (Arecaceae), a rare palm species restricted to one population in mist forests from Lord Howe Island, shows no negative genetic response to fragmentation (Shapcott *et al.*, 2011). In Australia, *Podocarpus elatus* (Podocarpaceae), a naturally discontinuous riparian species that has been further fragmented over 200 years, shows no negative genetic effects (Mellick *et al.*, 2011). The endemic and endangered Seychelles tree *Glionnetia sericea* (Rubiaceae) has a naturally scattered distribution in mist forests at high altitude. Despite forest fragmentation and drastic reductions in population sizes and numbers, leaving isolated populations, genetic data demonstrate that these populations are still connected by pollen and seed exchange (Finger, 2012). Nevertheless, contrasting examples exist where naturally restricted species show negative genetic effects after forest fragmentation. *Metrosideros bartlettii* (Myrtaceae) has only 31 surviving individuals in three remnant native forest patches in the far north of New Zealand and shows problems with recruitment, low genetic diversity and high genetic differentiation, despite being naturally fragmented (Drummond *et al.*, 2000).

Concluding from these examples, it seems that persistence in small populations and resistance to genetic degradation appear to be greater for species that have long been fragmented, or are rare or isolated, compared with species that have recently been reduced to small populations. Nevertheless, opposing examples underline the fact that many more aspects and ecological traits are important to determine potential negative genetic effects following habitat fragmentation.

Box 5.4. Historical species distribution and genetic response to fragmentation in animals

Similar effects to those observed in plants can be observed for animal species. A study on 230 golden-crowned sifakas (*Propithecus tattersalli*) in the Daraina region of northern Madagascar demonstrated a high resilience towards forest fragmentation (Quémére *et al.*, 2010). These animals were shown to use the matrix and even to cross unpaved roads. However, there is also evidence that the underlying fragmentation history of this region is rather old (*c*.5000–10000 years ago) and started in the course of post-Pleistocene warming and aridification, long before the arrival of man about 2000 years ago (Quémére *et al.*, 2012). A similar lack of fragmentation effects has been shown in birds (Wright *et al.*, 2005) and other mammals (Bowyer *et al.*, 2002; Reddy *et al.*, 2012). If fragmentation occurred slowly enough, natural fragmentation processes may have enabled these species to evolve adaptations and resilience to fragmentation.

periods might be less vulnerable to genetic fragmentation compared with more widespread species that have recently become highly fragmented.

Gene flow, genetic differentiation and mating systems

Long-distance gene flow can be important for maintaining genetic diversity in fragmented populations by preventing inbreeding and limiting genetic drift (Hartl and Clark, 1997), which may ensure the viability of populations in the long-term (Lande, 1988, 1995; Petit and Hampe, 2006; Kramer *et al.*, 2008). A recent simulation study explored mathematically which demographic and genetic processes may determine the long-term persistence of metapopulations (i.e. the viability of fragmented populations; Theodorou *et al.*, 2009). The authors showed that species with intermediate dispersal rates, high plasticity of dispersers to habitat and/or established mechanisms of inbreeding avoidance may avoid negative fragmentation effects best.

Dispersal rates should at least be more than one individual per generation per population, but should not be too high, as death rates in the matrix may otherwise exceed the reproductive output of populations. Gene flow also requires that the matrix is in general usable for migratory movements, and optimal dispersal rates and matrix quality interact with each other (Theodorou *et al.*, 2009). Viable metapopulation dynamics also require that individual migrants possess sufficiently large dispersal distances to bridge the matrix between fragments. If this is not the case, or if patch quality is deteriorating in parallel with fragmentation due to intensified habitat exploitation, migrants may not settle successfully and populations can then be expected to become genetically isolated and prone to genetic drift, inbreeding and genetic bottlenecks, all of which are well-established negative consequences of a small population size.

Genetic analyses can directly (parentage analysis) and indirectly (genetic relatedness between populations) test the extent to which gene flow may be disrupted by habitat fragmentation, and the implication of this on population viability can be explored (Frankham *et al.*, 2004). Individuals can be uniquely identified; reproductive success, mating patterns, kinship and fitness in natural populations can be investigated using molecular techniques, which contribute to a better understanding of population ecology (Queller *et al.*, 1993; Petrie and Kempenaers, 1998; Luikart and England, 1999; Sunnucks, 2000). The mating system of a species can be explored with genetic data, which can be important, as it can, at least in the short term, buffer against the negative effects of population fragmentation and isolation. Different mating strategies (such as outbreeding and self-compatibility) will diversely affect species' genetic structure and dispersal. The considerable difference between plant and animal species is the degree of mobility. While plant species are mostly restricted to dispersal through their pollen and seeds, animal species often actively disperse in forests. Therefore, differences can be expected in the effects of forest fragmentation on gene flow and genetic differentiation for different taxonomic groups (Boxes 5.5 and 5.6).

Methodological limits to detecting genetic consequences of forest fragmentation

While habitat loss seems to have obvious negative effects on biodiversity, and many studies demonstrate that reductions in numbers, small population sizes, genetic isolation and inbreeding have negative genetic effects on species, it seems to be difficult in some cases to detect fragmentation effects at all. The interpretation that some species may simply cope well with forest fragmentation could be a plausible interpretation in some cases. Nevertheless, other explanations such as methodological problems, such as sampling scheme, choice of molecular markers and long generation times (time lag between fragmentation events and the possibility to detect genetic effects), could lead to misinterpretations (Boxes 5.7 and 5.8).

Outlook

Importance of genetic research for understanding forest fragmentation

As demonstrated in the above sections, one of the main genetic threats resulting from forest fragmentation is genetic isolation in the resulting forest patches, the subsequent loss of genetic diversity and increased inbreeding for species that dwell in isolated forest islands. These genetic problems are expected to be largely detrimental for all taxonomic groups considered here. Besides genetic consequences, there are obvious negative ecological effects from fragmentation for forest species, such as loss of higher-order biodiversity and reduced effectiveness of key ecological processes, such as nutrient cycling (covered explicitly in Chapters 4 and 7, this volume). The interruption of potential changes in mutualistic and antagonistic plant–animal interactions may also affect genetic processes and thus offspring fitness (Aizen and Feinsinger, 1994; Tscharntke and Brandl, 2004; Valladares *et al.*, 2006). All these processes will determine the community structure within small forest fragments and

Box 5.5. Gene flow in plants in fragmented forest systems

Forest tree species have shown a wide range of responses to forest fragmentation (Kramer *et al.*, 2008; Bacles and Jump, 2011), with the vulnerabilities of species depending on several reproductive and demographic factors (Ghazoul, 2005). There is support in the literature for a fitness benefit for tree species with the ability for long-distance gene flow (Dick, 2001; Petit and Hampe, 2006; Sork and Smouse, 2006; Kramer *et al.*, 2008; Lander *et al.*, 2010). Further evidence is apparent in specific studies. For example, *Gomortega keule*, an endangered Chilean tree, exhibits long-distance pollen dispersal by insects, despite extremely small and highly fragmented populations (Lander *et al.*, 2010). Individual isolated trees within the matrix exchanged pollen over long distances, demonstrating the lack of genetic isolation in a highly fragmented landscape. However, the matrix type can be very important, with a potential negative impact on pollination success because pollinators become diverted to crop species by some land uses more than others (Lander *et al.*, 2011). No negative genetic effects of forest fragmentation were observed for the insect-pollinated and wind-dispersed *Entandrophragma cylindricum* (Meliaceae), having pollen flow distances of up to 2 km (Lourmas *et al.*, 2007). However, a lack of long-distance gene flow can lead to negative genetic effects. For example, comparing adult and juvenile genetic data of an endemic tree of the Seychelles, *Vateriopsis seychellarum*, it could be shown that a strong reduction in the number of adults led to an increase in genetic differentiation and population isolation in juveniles due to low gene flow rates (Finger *et al.*, 2012). These findings suggest that plant species' pollen-dispersal vectors that are able to disperse effectively over the matrix may be less vulnerable to negative genetic effects when forests are fragmented. However, the type of matrix is likely to be important.

Breeding system characteristics such as self-compatibility in plants may also contribute to the persistence of populations in isolation, even when gene flow is low (Levin, 2010). Selfing can, for example, have benefits in terms of predator avoidance (Ghazoul and Satake, 2009), reducing maladapted genes or assuring reproduction (Levin, 2010). Nevertheless, self-fertilization also leads to increased homozygosity (Hamrick and Godt, 1996; Ingvarsson, 2002) and selfed progeny tend to have higher rates of abortion or mortality at early ontogenic stages and shorter life history traits (Momose *et al.*, 1996; Ghazoul,1997; Ghazoul *et al.*, 1998; Naito *et al.*, 2005). Interestingly, Angeloni *et al.* (2011) suggested that the mating system of a plant species does not, in general, influence the magnitude of inbreeding depression, as it is found equally in both self-compatible and self-incompatible species. In a comparison of two dipterocarp species with contrasting mating systems, the outcrossing *Shorea leprosula* exhibited a greater loss of genetic diversity due to logging than the self-compatible *Shorea ovalis* (Ng *et al.*, 2009). Although logging reduces population density but not necessarily continuity, this result demonstrated that the mating systems of species may be important when populations are being reduced in size and increasingly fragmented. Thus, self-compatibility of species may be an important trait, allowing population persistence following fragmentation (Ng *et al.*, 2009; Levin, 2010). Whether such plant species will suffer from inbreeding depression at a later development stage or in later generations is difficult to predict (Husband and Schemske, 1996; Ingvarsson, 2002). Thus, it is important to investigate to what extent the long-term resilience of species to negative genetic and ecological effects of forest fragmentation can be linked to their mating systems.

which forest types will be able to harbour a high proportion of biodiversity. Indeed, Struebig *et al.* (2011) demonstrated that species richness and genetic diversity decline in parallel in insectivorous bat communities in a highly fragmented rainforest landscape (undergoing conversion to oil palm), highlighting the multidimensional scale at which biodiversity is affected by forest fragmentation.

Tree and plant species in general are not only threatened by habitat loss per se but often depend on forest-dwelling animal species for pollination and seed dispersal. Thus, particularly tree species, which lose a high proportion of their total population size due to logging, may further lose their potential for dispersal and successful pollination events in small forest fragments, as shown in Malaysia, where the lost megafauna led to reduced seed dispersal for large-seeded plants included in the study (Campos-Arceiz *et al.*, 2012). The long generation times of

Box 5.6. Gene flow in animals in fragmented forest systems

Many empirical studies have demonstrated a relationship between forest fragmentation, reduction of gene flow, and an increased level of genetic differentiation and genetic structure among animal populations. For example, in a study on two log-dwelling beetles with differential dispersal potential in south-eastern Australia, both showed reduced mobility and gene flow in fragmented compared with continuous forest, because the intervening matrix (pine plantation) posed a barrier to gene flow (Schmuki *et al.*, 2006). A recent study on the grey mouse lemur (*Microcebus murinus*) in north-western Madagascar reported a significantly increased degree of genetic differentiation, measured on the basis of mitochondrial DNA sequences (Schneider *et al.*, 2010) and nuclear microsatellites (U. Radespiel, unpublished results), among populations that were separated by open savannah compared with those that were separated by continuous forest. Similar fragmentation effects have even be detected in much more mobile animals, such as birds, as for example shown in the Australian logrunner (*Orthonyx temminckii*) (Pavlacky *et al.*, 2009). By comparing microsatellite genotypes of 220 logrunners from 11 sites in a partly deforested area in south-east Queensland, the authors showed that the proportion of deforested area correlated negatively with the migration rate of the birds. A recent study on seven forest-dependent bird species in a Kenyan cloud forest in the Eastern Arc Mountains (Callens *et al.*, 2011) showed that species that were more sedentary were more strongly differentiated than the more mobile species. This study also confirmed that landscape features such as low-altitude valleys influence historic genetic variation. However, it also became evident that some species underwent changes in mobility over time, demonstrating the importance of considering different timescales in the analyses of genetic structure and gene flow. As was discussed in Box 5.5, complex spatial patterns of genetic variation may only be fully understood if recent as well as historic population processes are addressed with appropriate molecular tools.

Animal species may develop different strategies to enable persistence in small populations. A study on Cunningham's skink (*Egernia cunninghami*) in New South Wales, Australia, showed that this species may avoid negative fragmentation effects by avoiding mating with close kin (Stow and Sunnucks, 2004). The authors showed that mates in fragments were less related to each other than by chance, thereby counteracting the effects of increasing relatedness in small populations. In contrast to plant species, sex-biased dispersal is a common characteristic for animal species and is thought to have evolved as a powerful inbreeding avoidance mechanism. A review on sex-biased dispersal of bird species has shown that for bird species, a high proportion of unpaired males are common in isolated populations or populations in small habitat fragments (Dale, 2003). Therefore, when females disperse, the vulnerability to extinction may increase in small or isolated populations. The author further predicted that migratory birds and short-lived species would be more threatened by extinction compared with resident birds and long-lived species due to the difference in time availability for females to locate unpaired males (Dale, 2003). Thus, following forest fragmentation, dispersal of the migrating sex may be prevented, or individuals may effectively be lost from the breeding population if they disperse into areas empty of conspecifics, resulting in disruption of dispersal routes and opportunities, which may lead to negative genetic effects and extinction for many animal species.

trees may be advantageous in that negative genetic effects may take longer to impact on species, but the drawback is that long-lived species are likely to have less potential to adapt quickly to changing environmental conditions. Species with narrow distributions, which depend on specific habitat conditions, may be further threatened, as they will not be able to adapt to the rapid environmental change associated with forest transformations, especially in the tropics. In Sabah (Malaysia), a study showed that 32 of 33 dipterocarp species are threatened and these are species with narrow distributions (Maycock *et al.*, 2012). Similarly, animal species that depend on undisturbed forest habitats, such as jaguars (Haag *et al.*, 2010), will not be able to survive in small fragmented forest patches due to their demanding habitat requirements.

Studies have started tackling genetic problems in small populations, such as genetic isolation and inbreeding depression, by applying genetic theories to experimental field or nursery experiments. This field of conservation genetics, and particularly the idea of genetic rescue, helps to inform conservation

Box 5.7. Avoiding type II errors in fragmentation genetics of plants

Diverse factors might explain why some studies conclude that, in fragmented populations of mature trees, genetic diversity has not diminished (Aguilar *et al.*, 2008). In particular, long-lived tree species may simply pre-date forest fragmentation. Consequently, signs of forest fragmentation influencing gene flow or mating systems may only be detected in offspring, and thus analysis of adults trees alone may lead to false-negative or type II errors. Indeed, Lowe *et al.* (2005) reviewed studies of several neotropical tree species and found mostly no effect on genetic diversity, but they did observe that, in the progeny, inbreeding, reproductive output and fitness were impacted. *Araucaria angustifolia* (Araucariaceae) from Brazil has experienced extensive habitat fragmentation since the 19th century and shows strong negative effects, such as reduced genetic diversity, increased levels of inbreeding and genetic differentiation among juvenile cohorts (Bittencourt and Sebbenn, 2009). In another Brazilian tree species, *Copaifera langsdorfii* (Caesalpinioideae), the progeny had negative genetic responses after 60–80 years of fragmentation (Sebbenn *et al.*, 2011). Interestingly, acute forest fragmentation in Europe has resulted in a decrease in diversity in adults of the European beech (*Fagus sylvatica*) in north-eastern Spain (Jump and Penuelas, 2006), a species that has certainly experienced several generations of forest fragmentation.

It is inevitably difficult to detect changes in genetic diversity for recently fragmented tree species in natural populations. In general, it is accepted that, following habitat fragmentation, the effects on genetic diversity have a lag time that is related to the generation time of the focal organism (Honnay *et al.*, 2005). Moreover, it may take several generations of isolation before the effects become apparent, creating a challenge for scientists to detect effects. Furthermore, genetic diversity is unlikely to be the only driver of fitness declines. In Japan, for example, *Magnolia obovata* (Magnoliaceae) had reduced genetic diversity yet showed no negative effects on reproductive output (Isagi *et al.*, 2007). Other ecological aspects may also be important. For example, it is suggested that the long-term effect of fragmentation on the genetic diversity of tropical trees depends on the amount of local forest cover in proximity to the fragmented populations (Pither *et al.*, 2003). These examples underline the need to study fragmentation at a multidimensional scale, ideally including analysis on genetic diversity, reproductive success, gene flow and offspring viability, as has been done in the case of *Eucalyptus socialis* in southern Australia (Breed *et al.*, 2012), suggesting that pollen diversity may have an even greater impact on fitness than inbreeding.

Box 5.8. Avoiding type II errors in fragmentation genetics of animals

As with plant systems, a meta-analysis on fragmentation effects in animal models confirmed that genetic fragmentation effects are not ubiquitous (Keyghobadi, 2007). Only 17 of 29 reviewed studies (58.6%) showed unambiguous fragmentation effects on genetic diversity. This relatively low number raises the question of why some species seem to be less susceptible to fragmentation than others. Potential explanations can be grouped into biologically meaningful explanations versus methodological artefacts. Genetic diversity of fragmented populations of animals may not be affected because: (i) their dispersal abilities allow them to cross the matrix; (ii) their generalist lifestyles may allow them to use habitat edges or the matrix effectively; (iii) they may use surrogate habitats; or (d) they may possess active mechanisms to avoid inbreeding (Theodorou *et al.*, 2009). Furthermore a reduction in genetic diversity might not be detected in fragmented populations of animals if: (i) the time since fragmentation was not long enough to produce molecular consequences given the species-typical generation time; (ii) the critical fragment size has not yet been reached beyond which the species starts to suffer genetic depletion; or (iii) the molecular markers used in the study are inappropriate for the detection of molecular changes at the temporal scale of the underlying fragmentation history (Radespiel and Bruford, 2014). In order to fully understand the interaction and relative importance of all factors involved, many more studies will be needed that tease these various explanations apart, not only for one species but for many different taxonomic groups.

and management of species or populations that survive in small numbers. Genetic rescue describes the increase in fitness of small populations resulting from the alleviation of inbreeding depression by immigrants (Thrall *et al.*, 1998). According to the genetic rescue hypothesis, immigrants can improve population fitness through their genetic contribution alone (Waite *et al.*, 2005). Examples of genetic rescue can be found for diverse organisms, showing that immigrant-mediated gene flow can significantly improve fitness in small inbred populations (Bryant *et al.*, 1999; Waite *et al.*, 2005; Hogg *et al.*, 2006). Indeed, inter-population pollination crosses from the large donor population to a small recipient population of the endemic and endangered tree *Medusagyne oppositifolia* in the Seychelles resulted in higher reproductive success relative to within-population crosses (Finger *et al.*, 2011). Nevertheless, the application of genetic rescue in practical conservation and management is rarely conducted due to costly and/or work intensive nursery or field trials. Future research should consider how to develop practical and general guidelines for conservation organizations to help their decision making on how to manage genetic biodiversity and species with small populations based on their taxonomic status, their life history, mating system and a rough understanding of gene flow in novel landscapes.

Similarly, *ex situ* conservation can benefit from genetic studies to screen the quality and quantity of collections. For example, *ex situ* collections have been able to capture 94% of the whole species neutral genetic diversity in a rare palm (*Leucothrinax morrisii*, Arecaceae) within 58 individuals (Namoff *et al.*, 2010), demonstrating that even a relatively modest number of individuals (100–200) can help conserve genetic diversity *ex situ*. Similarly, in the endangered endemic dipterocarp *Vatica guangxiensis* in China, 28 *ex situ* individuals conserved 88.31% of the neutral genetic variation of natural populations (Li *et al.*, 2002). Considering that target 8 of the Global Strategy for Plant Conservation is, by 2020, to have 'at least 75% of threatened plant species in *ex situ* collections, preferably in the country of origin, and at least 20% available for recovery and restoration programs', it is obvious that genetic information would be crucial to ensure that collections represent species' genetic diversities, promising the most effective protection and potential for restoration.

Finally, to make sensible restoration management plans, it is important to include genetic studies to ensure that restoration projects: (i) do not use inbred plant material to prevent inbreeding depression, e.g. due to wrong germplasm collection strategies (Ennos, 2003; Lengkeek *et al.*, 2005; Burgarella *et al.*, 2007); and (ii) prevent outbreeding depression, which can disrupt local adaptation (Hartl and Clark, 1997). The link between ecological and genetic processes that might contribute to species survival in natural and disturbed habitats remains poorly documented.

Priority areas for future studies

Genetic research demonstrates that generalizations about the negative genetic effects of forest fragmentation across species are extremely difficult. For example, even in a study of four co-occurring woody species from a single region in south-western Spain, contrasting life histories lead to very different genetic responses to the same fragmentation processes (Aparicio *et al.*, 2012). The challenge to disentangle causes for differing genetic and ecological effects to forest fragmentation is therefore great and will surely require interdisciplinary studies. For decision making in conservation, it will be useful to define, in particular, species traits that are most likely to infer negative fragmentation genetic responses, such as gene flow distances, mating system, pollination mode and distributional range. Tree species, for example, could then be categorized into thread levels and, for example, logging effects could be predicted at different scales. This is particularly important for tropical forest species, which currently face rapid and extensive fragmentation of their habitats.

As forest trees and forest plants are keystone species of forest ecosystems, their continued existence is essential for species' compositions and interactions in these ecosystems. The study of community genomics in the context of forest fragmentation is therefore of particular interest. Few studies exist that link forest fragmentation genetics to biodiversity conservation and ecosystem function. For example, we have little idea what the community consequences are of reductions of inter- and intra-specific genetic diversity in forest tree species for higher trophic level diversity. Are genetically depauperate forests less able to provide ecosystem services, such as carbon storage? Combining next-generation molecular-based approaches and ecological data, the implications of forest fragmentation for community-level processes could be better understood. In addition, advancing our understanding of adaptive variation will be especially important when it comes to forest restoration and species translocations. The preservation of forest fragments as well as the maintenance and creation of corridors to enhance gene flow between populations will clearly be important. However, not much is known about the effectiveness of corridors for many species and the different types of corridors needed for different species. One important challenge will be to estimate minimum viable population sizes, which will surely vary drastically among species, for species occurring in forest fragments, so that *in situ* and *ex situ* management for forests can be planned.

Finally, the transition from conservation genetics to conservation genomics will bring new challenges and opportunities, especially in terms of handling the amount of data and filtering the right information to help conservation of forest species. In particular, when it comes to predicting species' ability to adapt to changed environmental conditions or climate change, neutral markers will be of limited use. As the number of studies increase that combine both genetic and ecological information – including that of adaptive genes – predicting the implications of fragmentation for species' adaptive potential will become increasingly feasible. Ensuring that this information is effectively transferred between scientists and conservation organizations will require considerable effort, and scientists must enhance the link between scientific understanding and operational guidelines for conservation and landscape managers.

References

Aguilar, R., Ashworth, L., Galetto, L. and Aizen, M.A. (2006) Plant reproductive susceptibility to habitat fragmentation: review and synthesis through a meta-analysis. *Ecology Letters* 9, 968–980.

Aguilar, R., Quesada, M., Ashworth, L., Herrerias-Diego, Y. and Lobo, J. (2008) Genetic consequences of habitat fragmentation in plant populations: susceptible signals in plant traits and methodological approaches. *Molecular Ecology* 17, 5177–5188.

Aizen, M.A. and Feinsinger, P. (1994) Forest fragmentation, pollination and plant reproduction in a chaco dry forest, Argentina. *Ecology* 75, 330–351.

Allendorf, F.W. and Luikart, G. (2007) *Conservation and the Genetics of Populations*. Blackwell Publishing, Malden, Massachusetts.

Andersen, L.W., Fog, K. and Damgaard, C. (2004) Habitat fragmentation causes bottlenecks and inbreeding in the European tree frog (*Hyla arborea*). *Proceedings of The Royal Society B: Biological Sciences* 271, 1293–1302.

Angeloni, F., Ouborg, N.J. and Leimu, R. (2011) Meta-analysis on the association of population size and life history with inbreeding depression in plants. *Biological Conservation* 144, 35–43.

Aparicio, A., Hampe, A., Fernandez-Carrillo, L. and Albaladejo, R.G. (2012) Fragmentation and comparative genetic structure of four Mediterranean woody species: complex interactions between life history traits and the landscape context. *Diversity and Distribution* 18, 226–235.

Avise, J.C. (2000) *Phylogeography: the History and Formation of Species*. Harvard University Press, Cambridge, Massachusetts.

Bacles, C.F.E. and Jump, A.S. (2011) Taking a tree's perspective on forest fragmentation genetics. *Trends in Plant Science* 16, 13–18.

Bacles, C.F.E., Burczyk, J., Lowe, A.J. and Ennos, R.A. (2005) Historical and contemporary mating patterns in remnant populations of the forest tree *Fraxinus excelsior* L. *Evolution* 59, 979–990.

Bijlsma, R., Bundgaard, J. and Boerema, A.C. (2000) Does inbreeding affect the extinction risk of small populations?: predictions from *Drosophila*. *Journal of Evolutionary Biology* 13, 502–514.

Bittencourt, J.V.M. and Sebbenn, A.M. (2009) Genetic effects of forest fragmentation in high-density *Araucaria angustifolia* populations in Southern Brazil. *Tree Genetics and Genomes* 5, 573–582.

Bowyer, J., Newell, G. and Eldridge, M.B. (2002) Genetic effects of habitat contraction on Lumholtz's tree-kangaroo (*Dendrolagus lumholtzi*) in the Australian Wet Tropics. *Conservation Genetics* 3, 61–67.

Breed, M.F., Marklund, M.H.K., Ottewell, K.M., Gardner, M.G., Harris, J.B. and Lowe, A.J. (2012) Pollen diversity matters: revealing the neglected effect of pollen diversity on fitness in fragmented landscapes. *Molecular Ecology* 21, 5955–5968.

Brigham, C.A. (2003) Factors affecting persistence in formerly common and historically rare plants. In: *Population Viability in Plants: Conservation, Management, and Modeling of Rare Plants*. Springer, Berlin/Heidelberg, pp. 59–97.

Brown, L., Ramey, R., Tamburini, B. and Gavin, T. (2004) Population structure and mitochondrial DNA variation in sedentary Neotropical birds isolated by forest fragmentation. *Conservation Genetics* 5, 743–757.

Bryant, E.H., Vackus, V.L., Clark, M.E. and Reed, D.H. (1999) Experimental tests of captive breeding for endangered species. *Conservation Biology* 13, 1487–1496.

Burdon, R.D. (2001) Genetic diversity and disease resistance: some considerations for research, breeding, and deployment. *Canadian Journal of Forest Research* 31, 596–606.

Burgarella, C., Navascues, M., Soto, A., Lora, A. and Fici, S. (2007) Narrow genetic base in forest restoration with holm oak (*Quercus ilex* L.) in Sicily. *Annals of Forest Science* 64, 757–763.

Caizergues, A., Ratti, O., Helle, P., Rotelli, L., Ellison, L. and Rasplus, J.Y. (2003) Population genetic structure of male black grouse (*Tetrao tetrix* L.) in fragmented vs. continuous landscapes. *Molecular Ecology* 12, 2297–2305.

Callens, T., Galbusera, P., Matthysen, E., Durand, E.Y., Githiru, M., Huyghe, J.R. and Lens, L. (2011) Genetic signature of population fragmentation varies with mobility in seven bird species of a fragmented Kenyan cloud forest. *Molecular Ecology* 20, 1829–1844.

Campos-Arceiz, A., Traeholt, C., Jaffar, R., Santamaria, L. and Corlett, R.T. (2012) Asian tapirs are no elephants when it comes to seed dispersal. *Biotropica* 44, 220–227.

Cannon, C.H., Morley, R.J. and Bush, A.B.G. (2009) The current refugial rainforests of Sundaland are unrepresentative of their biogeographic past and highly vulnerable to disturbance. *Proceedings of the National Academy of Sciences USA* 106, 11188–11193.

Cascante, A., Quesada, M., Lobo, J.J. and Fuchs, E.A. (2002) Effects of dry tropical forest fragmentation on the reproductive success and genetic structure of the tree *Samanea saman*. *Conservation Biology* 16, 137–147.

Charpentier, M.J.E., Williams, C.V. and Drea, C.M. (2008) Inbreeding in ring-tailed lemurs (*Lemur catta*): genetic diversity predicts parasitism, immunocompetence, and survivorship. *Conservation Genetics*, 1605–1615.

Cole, C.T. (2003) Genetic variation in rare and common plants. *Annual Review of Ecology, Evolution, and Systematics* 34, 213–237.

Corlett, R.T. (2011) Impacts of warming on tropical lowland rainforests. *Trends in Ecology and Evolution* 26, 606–613.

Craul, M., Chikhi, L., Sousa, V., Olivien, G.L., Rabesandratana, A., Zimmermann, E. and Radespiel, U. (2009) Influence of forest fragmentation on an endangered large-bodied lemur in northwestern Madagascar. *Biological Conservation* 142, 2862–2871.

Cuartas-Hernandez, S. and Nunez-Farfan, J. (2006) The genetic structure of the tropical understorey herb *Dieffenbachia seguine* L. before and after forest fragmentation. *Evolutionary Ecology Research* 8, 1061–1075.

Dale, S. (2003) Female-biased dispersal, low female recruitment, unpaired males, and the extinction of small and isolated bird populations. *Oikos* 92, 344–356.

Dick, C.W. (2001) Genetic rescue of remnant tropical trees by an alien pollinator. *Proceedings of the Royal Society of London Series B: Biological Sciences* 268, 2391–2396.

Drummond, R.S.M., Keeling, D.J., Richardson, T.E., Gardner, R.C. and Wright, S.D. (2000) Genetic analysis and conservation of 31 surviving individuals of a rare New Zealand tree, *Metrosideros bartlettii* (Myrtaceae). *Molecular Ecology* 9, 1149–1157.

Eliades, N.-G.H., Gailing, O., Leinemann, L., Fady, B. and Finkeldey, R. (2011) High genetic diversity and significant population structure in *Cedrus brevifolia* Henry, a narrow endemic Mediterranean tree from Cyprus. *Plant Systematics and Evolution* 294, 185–198.

Ennos, R.A. (2003) The contribution of population genetic studies to plant conservation. *Botanical Journal of Scotland* 55, 89–100.
Evans, S.R. and Sheldon, B.C. (2008) Interspecific patterns of genetic diversity in birds: correlations with extinction risk. *Conservation Biology* 22, 1016–1025.
Fahrig, L. (2003) Effects of habitat fragmentation on biodiversity. *Annual Review of Ecology, Evolution, and Systematics* 34, 487–515.
Finger, A. (2012) Ecological and genetic processes underlying Allee effects among tree populations in the context of divergent population histories. PhD thesis, ETH Zurich, Switzerland.
Finger, A., Kettle, C.J., Kaiser-Bunbury, C.N., Valentin, T., Doudee, D., Matatiken, D. and Ghazoul, J. (2011) Back from the brink: potential for genetic rescue in a critically endangered tree. *Molecular Ecology* 20, 3773–3784.
Finger, A., Kettle, C.J., Kaiser-Bunbury, C.N., Valentin, T., Mougal, J. and Ghazoul, J. (2012) Forest fragmentation genetics in a formerly widespread island tree: *Vateriopsis seychellarum* (Dipterocarpaceae). *Molecular Ecology* 21, 2369–2382.
Finke, E. and Jetschke, G. (1999) How inbreeding and outbreeding influence the risk of extinction: a genetically explicit model. *Mathematical Biosciences* 156, 309–314.
Fowler, K. and Whitlock, M.C. (1999) The variance in inbreeding depression and the recovery of fitness in bottlenecked populations. *Proceedings of the Royal Society of London B* 266, 2061–2066.
Frankham, R. (1995) Inbreeding and extinction: a threshold effect. *Conservation Biology* 9, 792–799.
Frankham, R. (1996) Relationship of genetic variation to population size in wildlife. *Conservation Biology* 10, 1500–1508.
Frankham, R. (1999) Resolving conceptual issues in conservation genetics: the roles of laboratory species and meta-analyses. *Hereditas* 130, 195–201.
Frankham, R., Ballou, J.D. and Briscoe, D.A. (2004) *A Primer of Conservation Genetics.* Cambridge University Press Cambridge, UK.
Garner, A., Rachlow, J.L. and Hicks, J.F. (2005) Patterns of genetic diversity and its loss in mammalian populations. *Conservation Biology* 19, 1215–1221.
Geburek, T. and Konrad, H. (2008) Why the conservation of forest genetic resources has not worked. *Conservation Biology* 22, 267–274.
Ghazoul, J. (1997) The pollination and breeding system of *Dipterocarpus obtusifolius* (Dipterocarpaceae) in dry deciduous forests of Thailand. *Journal of Natural History* 31, 901–916.
Ghazoul, J. (2005) Pollen and seed dispersal among dispersed plants. *Biological Reviews* 80, 413–443.
Ghazoul, J. and Satake, A. (2009) Nonviable seed set enhances plant fitness: the sacrificial sibling hypothesis. *Ecology* 90, 369–377.
Ghazoul, J., Liston, K.A. and Boyle, T.J.B. (1998) Disturbance-induced density-dependent seed set in *Shorea siamensis* (Dipterocarpaceae), a tropical forest tree. *Journal of Ecology* 86, 462–473.
Glitzendanner, M.A. and Soltis, P.S. (2000) Patterns of genetic variation in rare and widespread plant congeners. *American Journal of Botany* 87, 783–792.
Goossens, B., Chikhi, L., Ancrenaz, M., Lackman-Ancrenaz, I., Andau, P. and Bruford, M.W. (2006) Genetic signature of anthropogenic population collapse in orang-utans. *PLoS Biology* 4, 285–291.
Haag, T., Santos, A.S., Sana, D.A., Morato, R.G., Cullen, L. Jr, Crawshaw, P.G. Jr, De Angelo, C., Di Bitetti, M.S., Salzano, F.M. and Eizirik, E. (2010) The effect of habitat fragmentation on the genetic structure of a top predator: loss of diversity and high differentiation among remnant populations of Atlantic Forest jaguars (*Panthera onca*). *Molecular Ecology* 19, 4906–4921.
Hamrick, J.L. (2004) Response of forest trees to global environmental changes. *Forest Ecology and Management* 197, 323–335.
Hamrick, J.L. and Godt, M.J.W. (1996) Effects of life history traits on genetic diversity in plant species. *Philosophical Transactions of the Royal Society of London Series B: Biological Sciences* 351, 1291–1298.
Hansen, M.C., Stehman, S.V. and Potapov, P.V. (2010) Quantification of global gross forest cover loss. *Proceedings of the National Academy of Sciences USA* 107, 8650–8655.
Hartl, D.L. and Clark, A.G. (1997) *Principles of Population Genetics*, 3rd edn. Sinauer Associates, Sunderland, Massachusetts.
Hirota, T., Hrohata, T., Mashima, H., Satoh, T. and Obara, Y. (2004) Population structure of the large Japanese field mouse, *Apodemus speciosus* (Rodentia: Muridae), in suburban landscape, based on mitochondrial D-loop sequences. *Molecular Ecology* 13, 3275–3282.
Hogg, J.T., Forbes, S.H., Steele, B.M. and Luikart, G. (2006) Genetic rescue of an insular population of large mammals. *Proceedings of The Royal Society B: Biological Sciences* 273, 1491–1499.

Honnay, O. and Jacquemyn, H. (2007) Susceptibility of common and rare plant species to the genetic consequences of habitat fragmentation. *Conservation Biology* 21, 823–831.

Honnay, O., Jacquemyn, H., Bossuyt, B. and Hermy, M. (2005) Forest fragmentation effects on patch occupancy and population viability of herbaceous plant species. *New Phytologist* 166, 723–736.

Husband, B.C. and Schemske, D.W. (1996) Evolution of the magnitude and timing of inbreeding depression in plants. *Evolution* 50, 54–70.

Ingvarsson, P.K. (2002) A metapopulation perspective on genetic diversity and differentiation in partially self-fertilizing plants. *Evolution* 56, 2368–2373.

Isagi, Y., Tateno, R., Matsuki, Y., Hirao, A., Watanabe, S. and Shibata, M. (2007) Genetic and reproductive consequences of forest fragmentation for populations of *Magnolia obovata*. *Ecological Research* 22, 382–389.

Ismail, S. (2013) Fragmentation genetics of tropical tree species in an agro-forest landscape. PhD thesis, ETH Zurich, Switzerland.

Ismail, S., Ghazoul, J., Ravikanth, G., Shaanker, R.U., Kushalappa, C.G. and Kettle, C.J. (2012) Does long distance pollen dispersal preclude inbreeding in tropical trees? Fragmentation genetics of *Dysoxylum malabaricum* in an agro-forest landscape. *Molecular Ecology* 21, 5484–5496.

Jablonski, D., Roy, K. and Valentine, J.W. (2006) Out of the tropics: evolutionary dynamics of the latitudinal diversity gradient. *Science* 314, 102–106.

Jump, A.S. and Penuelas, J. (2006) Genetic effects of chronic habitat fragmentation in a wind-pollinated tree. *Proceedings of the National Academy of Sciences USA* 103, 8096–8100.

Karron, J.D. (1991) Patterns of genetic variation and breeding systems in rare plant species. In: Falk, D.A. and Holsinger K.E. (eds) *Genetics and Conservation of Rare Plants*. Oxford University Press, New York, pp. 87–98.

Keller, L.F. and Waller, D.M. (2002) Inbreeding effects in wild populations. *Trends in Ecology and Evolution* 17, 230–241.

Keyghobadi, N. (2007) The genetic implications of habitat fragmentation for animals. *Canadian Journal of Zoology* 85, 1049–1064.

Kramer, A.T., Ison, J.L., Ashley, M.V. and Howe, H.F. (2008) The paradox of forest fragmentation genetics. *Conservation Biology* 22, 878–885.

Laikre, L., Allendorf, F.W., Aroner, L.C., Baker, C.S., Gregovich, D.P., Hansen, M.M., Jackson, J.A., Kendall, K.C., McKelvey, K., Neel, M.C., Olivieri, I., Ryman, N. *et al.* (2010) Neglect of genetic diversity in implementation of the convention on biological diversity. *Conservation Biology* 24, 86–88.

Lancaster, M.L., Taylor, A.C., Cooper, S.J.B. and Carthew, S.M. (2011) Limited ecological connectivity of an arboreal marsupial across a forest/plantation landscape despite apparent resilience to fragmentation. *Molecular Ecology* 20, 2258–2271.

Lande, R. (1988) Genetics and demography in biological conservation. *Science* 241, 1455–1460.

Lande, R. (1995) Mutation and conservation. *Conservation Biology* 9, 782–791.

Lande, R. and Schemske, D.W. (1985) The evolution of self-fertilization and inbreeding depression in plants. *Evolution* 39, 24–40.

Lander, T.A., Boshier, D.H. and Harris, S.A. (2010) Fragmented but not isolated: contribution of single trees, small patches and long-distance pollen flow to genetic connectivity for *Gomortega keule*, an endangered Chilean tree. *Biological Conservation* 143, 2583–2590.

Lander, T.A., Bebber, D.P., Choy, C.T.L., Harris, S.A. and Boshier, D.H. (2011) The Circe principle explains how resource-rich land can waylay pollinators in fragmented landscapes. *Current Biology* 21, 1302–1307.

Leimu, R. and Mutikainen, P. (2005) Population history, mating system, and fitness variation in a perennial herb with a fragmented distribution. *Conservation Biology* 19, 349–356.

Leimu, R., Mutikainen, P., Koricheva, J. and Fischer, M. (2006) How general are positive relationships between plant population size, fitness and genetic variation? *Journal of Ecology* 94, 942–952.

Lengkeek, A.G., Kindt, R., van der Maesen, L.J.G., Simons, A.J. and van Oijen, D.C.C. (2005) Tree density and germplasm source in agroforestry ecosystems in Meru, Mount Kenya. *Genetic Resources and Crop Evolution* 52, 709–721.

Levin, D.A. (2010) Environment-enhanced self-fertilization: implications for niche shifts in adjacent populations. *Journal of Ecology* 98, 1276–1283.

Li, Q., Xua, Z. and He, T. (2002) *Ex situ* genetic conservation of endangered *Vatica guangxiensis* (Dipterocarpaceae) in China. *Biological Conservation* 106, 151–156.

Loss, S.R., Terwilliger, L.A. and Peterson, A.C. (2011) Assisted colonization: integrating conservation strategies in the face of climate change. *Biological Conservation* 144, 92–100.

Lourmas, M., Kjellberg, F., Dessard, H., Joly, H.I. and Chevallier, M.H. (2007) Reduced density due to logging and its consequences on mating system and pollen flow in the African mahogany *Entandrophragma cylindricum*. *Heredity* 99, 151–160.

Lowe, A.J., Boshier, D., Ward, M., Bacles, C.F.E. and Navarro, C. (2005) Genetic resource impacts of habitat loss and degradation; reconciling empirical evidence and predicted theory for Neotropical trees. *Heredity* 95, 255–273.

Luikart, G. and England, P.R. (1999) Statistical analysis of microsatellite DNA data. *Trends in Ecology and Evolution* 14, 253–256.

Malone, C., Knapp, C., Taylor, J. and Davis, S. (2003) Genetic consequences of Pleistocene fragmentation: isolation, drift, and loss of diversity in rock iguanas (*Cyclura*). *Conservation Genetics* 4, 1–15.

Mansourian, S., Vallauri, D. and Dudley, N. (2005) *Forest Restoration in Landscapes: Beyond Planting Trees*. Springer, New York.

Maycock, C.R., Kettle, C.J., Khoo, E., Pereira, J.T., Sugau, J.B., Nilus, R., Ong, R.C., Amaludin, N.A., Newman, M.F. and Burslem, F.R.P. (2012) A revised conservation assessment of dipterocarps in Sabah. *Biotropica* 44, 649–657.

McConkey, K.R., Prasad, S., Corlett, R.T., Campos-Arceiz, A., Brodie, J.F., Rogers, H. and Santamaria, L. (2012) Seed dispersal in changing landscapes. *Biological Conservation* 146, 1–13.

McNeely, J.A., Miller, K.R., Reid, W.V., Mittermeier, R.A. and Werner, T.B. (1990) *Conserving the World's Biological Diversity*. Washington, D.C., IUCN, World Resources Institute, Conservation International, WWFUS and the World Bank.

Mellick, R., Lowe, A. and Rossetto, M. (2011) Consequences of long- and short-term fragmentation on the genetic diversity and differentiation of a late successional rainforest conifer. *Australian Journal of Botany* 59, 351–362.

Mills, L.S. and Smouse, P.E. (1994) Demographic consequences of inbreeding in remnant populations. *American Naturalist* 144, 412–431.

Momose, K., Nagamitsu, T. and Inoue, T. (1996) The reproductive ecology of an emergent dipterocarp in a lowland rain forest in Sarawak. *Plant Species Biology* 11, 189–198.

Moritz, C. (1995) Uses of molecular phylogenies for conservation. *Philosophical Transactions of The Royal Society B: Biological Sciences* 349, 113–118.

Naito, Y., Konuma, A., Iwata, H., Suyama, Y., Seiwa, K., Okuda, T., Lee, S.L., Muhammad, N. and Tsumura, Y. (2005) Selfing and inbreeding depression in seeds and seedlings of *Neobalanocarpus heimii* (Dipterocarpaceae). *Journal of Plant Research* 118, 423–430.

Namkoong, G. (1991) Maintaining genetic diversity in breeding for resistance in forest trees. *Annual Review of Phytopathology* 29, 325–342.

Namoff, S., Husby, C.E., Francisco-Ortega, J., Noblick, L.R., Lewsi, C.C. and Griffith, P. (2010) How well does a botanical garden collection of a rare palm capture the genetic variation in a wild population? *Biological Conservation* 143, 1110–1117.

Newman, D. and Pilson, D. (1997) Increased probability of extinction due to decreased genetic effective population size: experimental populations of *Clarkia pulchella*. *Evolution* 51, 354–362.

Ng, K.K.S., Lee, S.L. and Ueno, S. (2009) Impact of selective logging on genetic diversity of two tropical tree species with contrasting breeding systems using direct comparison and simulation methods. *Forest Ecology and Management* 257, 107–116.

Nieminen, M., Singer, M.C., Fortelius, W., Schops, K. and Hanski, I. (2001) Experimental confirmation that inbreeding depression increases extinction risk in butterfly populations. *American Naturalist* 157, 237–244.

O'Grady, J.J., Burgman, M.A., Keith, D.A., Master, L.L., Adelman, S.J., Brook, B.W., Hammerson, G.A., Regan, T. and Frankham, R. (2004) Correlations among extinction risks assessed by different systems of threatened species categorization. *Conservation Biology* 18, 1624–1635.

Oliveras de Ita, A., Oyama, K., Smith, T.B., Wayne, R.K. and Mila, B. (2012) Genetic evidence for recent range fragmentation and severely restricted dispersal in the critically endangered Sierra Madre Sparrow, *Xenospiza baileyi*. *Conservation Genetics* 13, 283–291.

Olivieri, G.L., Sousa, V., Chikhi, L. and Radespiel, U. (2008) From genetic diversity and structure to conservation: genetic signature of recent population declines in three mouse lemur species (*Microcebus* spp.). *Biological Conservation*, 1257–1271.

Pavlacky, D.C., Goldizen, A.W., Prentis, P.J., Nicholls, J.A. and Lowe, A.J. (2009) A landscape genetics approach for quantifying the relative influence of historic and contemporary habitat heterogeneity on the genetic connectivity of a rainforest bird. *Molecular Ecology* 18, 2945–2960.

Petit, R.J. and Hampe, A. (2006) Some evolutionary consequences of being a tree. *Annual Review of Ecology, Evolution, and Systematics* 37, 187–214.

Petrie, M. and Kempenaers, B. (1998) Extra-pair paternity in birds: explaining variations between species and populations. *Trends in Ecology and Evolution* 13, 52–58.

Pither, R., Shore, J.S. and Kellman, M. (2003) Genetic diversity of the tropical tree *Terminalia amazonia* (Combretaceae) in naturally fragmented populations. *Heredity* 91, 307–313.

Potter, S., Eldridge, M.B., Cooper, S.B., Paplinska, J. and Taggart, D. (2012) Habitat connectivity, more than species' biology, influences genetic differentiation in a habitat specialist, the short-eared rock-wallaby (*Petrogale brachyotis*). *Conservation Genetics*, 937–952.

Queller, D.C., Strassmann, J.E. and Hugues, C.R. (1993) Microsatellites and kinship. *Trends in Ecology and Evolution* 8, 285–288.

Quémére, E., Crouau-Roy, B., Rabarivola, C., Louis, E.E. and Chikhi, L. (2010) Landscape genetics of an endangered lemur (*Propithecus tattersalli*) within its entire fragmented range. *Molecular Ecology* 19, 1606–1621.

Quémére, E., Amelot, X., Pierson, J., Crouau-Roy, B. and Chikhi, L. (2012) Genetic data suggest a natural prehuman origin of open habitats in northern Madagascar and question the deforestation narrative in this region. *Proceedings of the National Academy of Sciences USA* 109, 13028–13033.

Radespiel, U. and Bruford, M.W. (2014) Fragmentation genetics of rainforest animals: insights from recent studies. *Conservation Genetics* 15, 245–260.

Reddy, P.A., Gour, D.S., Bhavanishankar, M., Jaggi, K., Hussain, S.M., Harika, K. and Shivaji, S. (2012) Genetic evidence of tiger population structure and migration within an isolated and fragmented landscape in Northwest India. *PLoS ONE* 7, e29827.

Reed, D.H. (2005) Relationship between population size and fitness. *Conservation Biology* 19, 563–568.

Reed, D.H. and Bryant, E.H. (2000) Experimental tests of minimum viable population size. *Animal Conservation* 3, 7–14.

Reed, D.H. and Frankham, R. (2003) Correlation between fitness and genetic diversity. *Conservation Biology* 17, 230–237.

Reusch, T.B.H., Ehlers, A., Hammerli, A. and Worm, B. (2005) Ecosystem recovery after climatic extremes enhanced by genotypic diversity. *Proceedings of the National Academy of Sciences USA* 102, 2826–2831.

Ross, K.A., Fox, B.J. and Fox, M.D. (2002) Changes to plant species richness in forest fragments: fragment age, disturbance and fire history may be as important as area. *Journal of Biogeography* 29, 749–765.

Saccheri, I., Kuussaari, M., Kankare, M., Vikman, P., Fortelius, W. and Hanski, I. (1998) Inbreeding and extinction in a butterfly metapopulation. *Nature* 392, 491–494.

Schmuki, C., Vorburger, C., Runciman, D., Maceachern, S. and Sunnucks, P. (2006) When log-dwellers meet loggers: impacts of forest fragmentation on two endemic log-dwelling beetles in southeastern Australia. *Molecular Ecology* 15, 1481–1492.

Schneider, N., Chikhi, L., Currant, M. and Radespiel, U. (2010) Signals of recent spatial expansions in the grey mouse lemur (*Microcebus murinus*). *BMC Evolutionary Biology* 10, 105.

Sebbenn, A.M., Carvalho, A.C., Freitas, M.L., Moraes, S.M., Gaino, A.P., da Silva, J.M., Jolivet, C. and Moraes, M.L. (2011) Low levels of realized seed and pollen gene flow and strong spatial genetic structure in a small, isolated and fragmented population of the tropical tree *Copaifera langsdorffii* Desf. *Heredity* 106, 134–145.

Shapcott, A., Hutton, I., Baker, W.J. and Auld, T.D. (2011) Conservation genetics and ecology of an endemic montane palm on Lord Howe Island and its potential for resilience. *Conservation Genetics* 13, 257–270.

Sommer, S. (2003) Effects of habitat fragmentation and changes of dispersal behaviour after a recent population decline on the genetic variability of noncoding and coding DNA of a monogamous Malagasy rodent. *Molecular Ecology* 12, 2845–2851.

Sork, V.L. and Smouse, P.E. (2006) Genetic analysis of landscape connectivity in tree populations. *Landscape Ecology* 21, 821–836.

Soulé, M.E. (1976) Allozyme variation, its determinants in space and time. In: Ayala, F.J. (ed.) *Molecular Evolution*. Sinauer Associates, Sunderland, Massachusetts, pp. 60–77.

Spielman, D., Brook, B.W., Briscoe, D.A. and Frankham, R. (2004) Does inbreeding and loss of genetic diversity decrease disease resistance? *Conservation Genetics* 5, 439–448.

Stephens, P.A., Sutherland, W.J. and Freckleton, R.P. (1999) What is the Allee effect? *Oikos* 87, 185–190.

Stow, A.J. and Sunnucks, P. (2004) Inbreeding avoidance in Cunningham's skinks (*Egernia cunninghami*) in natural and fragmented habitat. *Molecular Ecology*, 443–447.

Struebig, M.J., Kingston, T., Petit, E.J., Le Comber, S.C., Zubaid, A., Mohd-Adnan, A. and Rossiter, S.J. (2011) Parallel declines in species and genetic diversity in tropical forest fragments. *Ecology Letters* 14, 582–590.

Sunnucks, P. (2000) Efficient genetic markers for population biology. *Trends in Ecology and Evolution* 15, 199–203.

Tanaka, Y. (2000) Extinction of populations by inbreeding depression under stochastic environments. *Population Ecology* 42, 55–62.

Theodorou, K., Souan, H. and Couvet, D. (2009) Metapopulation persistence in fragmented landscapes: significant interactions between genetic and demographic processes. *Journal of Evolutionary Biology* 22, 152–162.

Thrall, P.H., Richards, C.M., McCauley, D.E. and Antonovics, J. (1998) Metapopulation collapse: the consequences of limited gene flow in spacially structured populations. In: Bascompte, J. and Sole, R.V. (eds) *Modeling Spatiotemporal Dynamics in Ecology*. Springer, Berlin, pp. 83–104.

Tscharntke, T. and Brandl, R. (2004) Plant–insect interactions in fragmented landscapes. *Annual Review of Entomology* 49, 405–430.

Valladares, G., Salvo, A. and Cagnolo, L. (2006) Habitat fragmentation effects on trophic processes of insect–plant food webs. *Conservation Biology* 20, 212–217.

Van Geert, A., Van Rossum, F. and Triest, L. (2008) Genetic diversity in adult and seedling populations of *Primula vulgaris* in a fragmented agricultural landscape. *Conservation Genetics* 9, 845–853.

Vandepitte, K., Roldan-Ruiz, I., Leus, L., Jacquemyn, H. and Honnay, O. (2009) Canopy closure shapes clonal diversity and fine-scale genetic structure in the dioecious understorey perennial *Mercurialis perennis*. *Journal of Ecology* 97, 404–414.

Wade, T.G., Riitters, K.H., Wickham, J.D. and Jones, K.B. (2003) Distribution and causes of global forest fragmentation. *Conservation Ecology* 7, 7.

Wahbe, T.R., Ritland, C., Bunnell, F.L. and Ritland, K. (2005) Population genetic structure of tailed frogs (*Ascaphus truei*) in clearcut and old-growth stream habitats in south coastal British Columbia. *Canadian Journal of Zoology* 83, 1460–1468.

Waite, T.A., Vucetich, J.A., Saurer, T., Kroninger, M. and Vaughn, E. (2005) Minimizing extinction risk through genetic rescue. *Animal Biodiversity and Conservation* 28, 121–130.

Ward, M., Dick, C.W., Gribel, R. and Lowe, A.J. (2005) To self, or not to self… a review of outcrossing and pollen-mediated gene flow in neotropical trees. *Heredity* 95, 246–254.

Woltmann, S., Kreiser, B. and Sherry, T. (2012) Fine-scale genetic population structure of an understory rainforest bird in Costa Rica. *Conservation Genetics* 13, 925–935.

Wright, S.J. and Muller-Landau, H.C. (2006) The future of tropical forest species. *Biotropica* 38, 287–301.

Wright, T.F., Rodriguez, A.M. and Fleischer, R.C. (2005) Vocal dialects, sex-biased dispersal, and microsatellite population structure in the parrot *Amazona auropalliata*. *Molecular Ecology* 14, 1197–1205.

Young, A., Boyle, T. and Brown, T. (1996) The population genetic consequences of habitat fragmentation for plants. *Trends in Ecology and Evolution* 11, 413–418.

6 Forest Fragmentation and Climate Change

Richard T. Corlett*
Xishuangbanna Tropical Botanical Garden, Chinese Academy of Sciences, Yunnan, PR China

Abstract

Although the net contribution of global land-use change to anthropogenic carbon dioxide emissions is relatively small (*c*.10%), gross emissions and sinks are large, particularly in the tropics, and offer globally significant opportunities for mitigation. Forest fragments retain carbon that would otherwise be lost, but their proportional importance to regional carbon stores increases as the total forest area, and thus the absolute contribution to global stores, declines. Moreover, isolated fragments tend to lose carbon through fires and edge-related mortality, and possibly also by liana growth and the loss of large-bodied dispersal agents. Forest fragments also affect climate through their biophysical impacts on land-surface processes and, although the global impact of these is relatively minor, the local impacts may be significant. Fragments themselves are vulnerable to climate change, both directly and through changes in fire frequency and intensity, while the gaps between fragments will reduce the ability of forest species to track climate change by movements along climate gradients. Protecting and restoring fragments can help mitigate climate change, by reducing emissions and increasing sinks, but is not necessarily the optimum use of limited resources if larger forest areas are available. If fragments are actively restored, it makes sense to use plants adapted to the expected future climate, even if this involves the 'managed translocation' of species outside their current ranges. While REDD+ could potentially provide funding for fragment restoration, fragments are also particularly suited to local conservation initiatives.

Introduction: The Contribution of Deforestation to Climate Change

As a result of the continuing explosive growth of carbon dioxide emissions from fossil fuels, the proportional contribution of global land-use change to total anthropogenic emissions has continued to decline, from around 20% in 1980 to perhaps 10% today (Houghton, 2012). These are net emissions, however, and these numbers hide much larger gross emissions (from losses of biomass and soil carbon) and gross sinks (in regrowth and intact vegetation), offering many opportunities for reducing emissions or increasing sinks through improved land-use management. The fairly steady net emissions from global land-use change over the last 50 years also hide a decline in emissions from the temperate zone and an increase in the tropics (Richter and Houghton, 2011). Gross emissions from forest loss and degradation in the tropics over the past decade have been around one-third

*corlett@xtbg.org.cn

of total anthropogenic carbon emissions, showing the potential for significantly reducing global emissions by improving tropical forest management (Richter and Houghton, 2011).

The international carbon-payment programme REDD (reducing emissions from deforestation and forest degradation) seeks to achieve this by providing financial incentives for developing countries to reduce deforestation and forest degradation (Hiraldo and Tanner, 2011). REDD+ adds incentives for conservation, sustainable management and the enhancement of forest carbon stocks, so that countries with low rates of deforestation and degradation can also participate. However, it is still not clear exactly which activities will be rewarded under the future REDD+ mechanism. Moreover, whatever these activities are, the large uncertainties in most of the figures quoted above – uncertainties that increase greatly as we move towards finer spatial scales – will need to be reduced before REDD+ can reach its full potential. The uncertainties increase further and are more difficult to reduce as we move into the future, with the impact of climate change on forest carbon sinks a large unknown (Reich, 2011). There are also potential mismatches between conservation targeted at carbon and at biodiversity, with the risk that a narrow focus on carbon will displace threats from high carbon forests with relatively low biodiversity to high biodiversity forests with lower carbon (Harrison and Paoli, 2012).

Land-use change is not just a source of carbon dioxide but also contributes to climate change through its impacts on methane and nitrous oxide emissions (Stocker *et al.*, 2013), surface albedo (the fraction of solar radiation reflected back into space) and evapotranspiration. Deforestation generally increases albedo, with a cooling effect on climate, but also reduces evapotranspiration, with a warming effect (Anderson-Teixeira *et al.*, 2012). In the tropics, the evapotranspiration effect tends to dominate, so deforestation has a net warming effect through these biophysical processes, in addition to the warming impact of carbon dioxide emissions. Crops differ considerably in albedo, however, and this will usually also vary over the cropping cycle, so the local climatic impacts will also vary in space and time (Georgescu *et al.*, 2013). In the boreal forest, in contrast, the albedo effect of deforestation is larger, because canopy loss exposes snow-covered surfaces, while the evapotranspiration effect is smaller, because evapotranspiration is lower, so deforestation has a net cooling effect through biophysical processes, which more or less cancels out the impact of the carbon dioxide released (Anderson-Teixeira *et al.*, 2012). In temperate forests, the albedo effect generally outweighs the evapotranspiration effect, but greenhouse gas effects are larger than both. On a global scale, the greenhouse gas effects of land-use change will dominate over the next century, but locally and immediately, the biophysical effects of deforestation will often be larger, particularly in the tropics. Indeed, large-scale forest clearance in the tropics can have a significant biophysical impact on air temperatures at the regional level (Hallgren *et al.*, 2013).

Biophysical effects can also have a large impact on rainfall, particularly in the tropics. Deforestation can lead to local increases in rainfall as a result of enhanced convection from surface heating, but reduces downwind rainfall on a regional scale (100–1000 km) as a result of reduced evapotranspiration (Aragão, 2012; Spracklen *et al.*, 2012). Moreover, deforestation fires produce aerosols that can reduce surface temperatures while warming the atmosphere, significantly reducing rainfall in a narrow band near the equator (Tosca *et al.*, 2010, 2012).

Note that the focus of REDD+ on greenhouse gases ignores the biophysical effects of deforestation. This will not matter if REDD+ deals only with tropical forests, where biophysical effects reinforce the warming impacts of carbon emissions, but it does mean that carbon alone cannot be used as a currency for comparing the climate impacts of deforestation in different climate zones.

Role of Forest Fragments as Stores, Sinks and Sources of Carbon

Deforestation is rarely complete, at least initially, with forest fragments of various sizes and shapes, and modified to varying extents, left uncleared for a huge variety of reasons. These fragments retain carbon in biomass and soils that would otherwise be released during or after the clearance process. Only the largest fragments will be accounted for in satellite surveys of forest cover so fragments of 'sub-pixel' size may be under-recorded stores of carbon. The significance of forest fragments to regional carbon budgets will depend on their total extent, their carbon density and their ability to retain forest structure and function over time, as well as the carbon density of alternative land uses. Unfortunately, there has been no global survey of forest fragments, and regional surveys vary greatly in their spatial resolution, their definition of fragmentation and the way in which their estimates are derived.

At one extreme, in the Brazilian Atlantic Forest region, forest fragments make up a substantial fraction of all remaining forest cover, with half the forest in patches <500 ha in area, 30% in patches <100 ha and 20% in patches <50 ha (Ribeiro *et al.*, 2009). However, although small fragments are an important fraction of the remaining forest cover, only 12% of the original forest remains. The fragments thus retain relatively little of the biomass carbon stored in the original landscape. On the other hand, where the deforested area is occupied by low-carbon pasture or annual crops (<10% of the carbon density of forest), these fragments may still store more biomass carbon than all other land uses combined (based on biomass estimates in Fearnside *et al.*, 2009). At the other extreme, in the Amazon basin, more than half the remaining forest is in patches >3,500,000 ha (35,000 km^2) and the numerous tiny fragments <400 ha in area contribute very little to the total forest area and carbon storage (Broadbent *et al.*, 2008). These two examples illustrate what appears to be a general pattern, with fragments increasingly important as *proportional* contributors to regional carbon stores as the total forest area remaining – and thus their *absolute* contribution to global stores – declines.

Crude percentages of the land area covered by fragments will often exaggerate their significance as biomass carbon stores. The fragments of forest left behind during deforestation are not a random sample of the original forest cover. In many cases, they will represent the lower carbon end of the variation in the landscape, such as rocky outcrops, swampy areas or other relatively infertile substrates. Often carbon storage will have been reduced further by collateral damage from cutting and fires during the deforestation process and/or the harvesting of valuable timber. Fires continue to threaten the carbon content of tropical forest fragments after isolation, escaping from adjacent cleared areas and burning into forest edges made vulnerable by lower humidity (Soares-Filho *et al.*, 2012). Fires then breed fires, by opening the canopy and increasing fuel availability. This is true also at the landscape scale, with more and more intense fires occurring in areas where the forest is more fragmented (Armenteras *et al.*, 2013). In the Brazilian Amazon, native or exotic grasses invade after canopy thinning, increasing fire intensity and potentially leading to 'savannization' of the forest (Silvério *et al.*, 2013). Note, however, that in more fire-prone forest types, such as eucalyptus and boreal forests, the non-forest matrix may actually reduce fire frequency and extent by acting as a fire break (Ross *et al.*, 2002; Senici *et al.*, 2010).

In highly dynamic landscapes, total fragment area is a balance between forest loss and forest regeneration. In a recently fragmented landscape, the losses will be largely old-growth forest and the gains young secondary forest, storing much less carbon in biomass. Over time, both the fragments and the landscape as a whole will become a mosaic of forests of different ages and biomasses (Lira *et al.*, 2012). Secondary forests may take >100–200 years to reach the biomass of primary forests (Mascaro *et al.*, 2012), so the continued loss of primary forest from fragmented landscapes may be reducing carbon storage even where the total forest cover is stable or increasing.

Even where fragments are left alone after isolation, increased edge mortality of trees as a result of desiccation and other edge processes can lead to a loss of biomass as large, shade-tolerant, old-growth tree species are replaced by pioneers (Laurance *et al.*, 2006; Pütz *et al.*, 2011; Tabarelli *et al.*, 2012). In the smallest fragments (<25 ha), biomass losses from edge processes may exceed 50% (Pütz *et al.*, 2011). It has been estimated that 2.6–4.5% of deforestation-related emissions in the Brazilian Amazon come from edge-related biomass loss (Numata *et al.*, 2010) and the proportion is likely to be considerably higher in more finely fragmented regions, such as the Brazilian Atlantic Forest. The generality of these processes still needs to be confirmed, however, as the evidence is currently largely from the continental Neotropics. Ramjohn *et al.* (2012) suggested that dry forest fragments on the hurricane-prone island of Puerto Rico may be more resistant to edge-related mortality. In some forests, the sealing of forest margins as edge-adapted species mature may serve to reduce the penetration distance of edge effects (Wicklein *et al.*, 2012), although the edge maturation process may be constrained by the management intensity of the adjacent matrix (Chabrerie *et al.*, 2013). Some fragments may be increasing in biomass, because of recovery from damage associated with clearance and/or the general carbon-fertilization effect on forests (Huntingford *et al.*, 2013).

Another potential mechanism for biomass carbon loss from fragments is the proliferation of lianas (Schnitzer and Bongers, 2011). Lianas reduce carbon storage if their leaves displace those of trees, as lianas store much less carbon per unit leaf area. It is not clear if this is occurring in fragment edges, where lianas may be of net benefit to fragment integrity by sealing gaps, but Jansen *et al.* (2010) suggested that the loss of large-bodied vertebrates from fragments (and other forests) may favour lianas, because they are predominantly wind dispersed and thus threaten carbon stocks. Again the evidence is largely neotropical and it is not clear what happens in South-east Asian dipterocarp forests, where the largest trees are wind dispersed.

More generally, some of the biotic changes in isolated fragments outlined elsewhere in this book (see Chapters 4 and 8, this volume) are likely to impact on the role of these fragments as carbon sinks and sources, perhaps in unpredictable ways. Brodie and Gibbs (2009) suggested that the loss of dispersal services from large-bodied vertebrates may threaten biomass carbon stocks by favouring smaller-seeded species with lower wood density, but this conclusion is based on a chain of rather weak links (Jansen *et al.*, 2010). Harrison *et al.* (2013) showed that the defaunation of the 52 ha forest dynamics plot at Lambir, Sarawak, in Malaysia, has resulted in pervasive changes in the tree community but found no evidence for an impact on biomass, at least in part because so much of this biomass is in wind-dispersed trees. Given the idiosyncrasies of forest tree biology, local effects could be large, but strong generalizations may be impossible.

Soil carbon is less vulnerable to deforestation than biomass carbon, but poor data quality (including insufficient sampling depth) limits our ability to make generalizations. A recent meta-analysis of the impacts of tropical land-use change showed that primary forest soils store most carbon, followed by secondary forests (–9%), grassland (–12%), cropland (–25%) and perennial crops (–30%) (Don *et al.*, 2011). Carbon loss after deforestation occurs mostly in the first few years, and soils in wetter regions appear to lose more carbon than those in drier regions. The authors note, however, that data coverage is still poor in the deforestation hotspots of tropical Asia and Africa. Global analyses (Guo and Gifford, 2002) have suggested that broadly similar patterns of carbon loss will occur outside the tropics, although data quality issues have still to be resolved (Smith *et al.*, 2012). Changes in soil carbon are not likely to be a big issue within forest fragments, unless there is a wholesale conversion of primary to secondary forest, but they become highly significant for carbon accounting at the landscape level, as most non-forest land uses store more carbon in the soil than biomass.

Overall, the evidence suggests that, while forest fragments may make a significant contribution to carbon storage in largely deforested regions, their global importance is fairly minor. The relative contribution of fragments to carbon fluxes is probably greater, however, because of the greater dynamism of forest near edges.

Biophysical Impacts of Forest Fragments on Climate

As highlighted in the introduction, deforestation affects climate through both its biogeochemical impacts on atmospheric composition and its biophysical impacts on land-surface processes. The biophysical impacts of forest fragments have not been investigated specifically but can be deduced, to a first approximation, from physical principles. In general, the impact of forest fragments on the radiation budget is expected to be proportional to the total area of forest remaining, with forest fragments usually standing out as dark (low albedo) patches in the brighter deforested landscape. Canopy disturbances, edge-related biomass loss and the replacement of primary forest by pioneers may change the albedo of forest patches, but there is little published data on this. In the eastern Amazon basin, young secondary forest had a higher albedo than primary forest, but after 10 years they were indistinguishable (Giambelluca *et al.*, 1997). Successional changes in albedo can be larger in boreal forests. In Alaska, the deciduous angiosperms favoured by severe fires have a significantly higher growing-season albedo than evergreen conifers (Beck *et al.*, 2011), while in Siberia, warming climates are driving the replacement of deciduous larch by evergreen conifers, with the largest impact on winter albedo when evergreen trees mask the snow (Shuman *et al.*, 2011).

The available data suggest that disturbances to forest fragments will typically increase their albedo and thus reduce their albedo-mediated warming influence. This biophysical warming influence will be partly (at high latitudes) or completely (in the tropics) outweighed by the cooling influence of evapotranspiration (Andersen-Teixeira *et al.*, 2012), but this cooling is also likely to be significantly reduced by increased water stress near forest margins (Briant *et al.*, 2010). The net result of these changes is not certain, but tropical forest fragments probably have a net cooling influence as long as water is available, while boreal fragments cause net warming. From a global perspective, these effects will be small in comparison with the carbon effects discussed above, but the impact on microclimates in and around the fragments may be biologically significant.

Impacts of Climate Change on Fragments

Climate change will impact directly on the carbon budgets of forest fragments through plant growth and mortality. The high edge-to-area ratio of small fragments makes them particularly vulnerable to any increase in the frequency and/or intensity of droughts (Briant *et al.*, 2010). Unfortunately, there is still considerable disagreement about the future trends in droughts among climate models, and the IPCC (2012) expressed only medium confidence in predictions that droughts will intensify in the 21st century in southern Europe and the Mediterranean, central Europe, central North America, Central America and Mexico, northeast Brazil and southern Africa, and low confidence in projections elsewhere. In practice, this means that the best we can say at present is, *if* droughts intensify *then* fragments will be vulnerable. Temperature projections are much more robust, with all models predicting an increase. Temperature increases may be higher in fragments because an influx of warm air from the surrounding open areas may overwhelm transpirational cooling by the canopy. Both rainfall declines and temperature increases are likely to increase tree mortality near edges, enhancing the edge-related biomass losses described above.

Climate change will also have indirect impacts through fire frequency and intensity. Again the large uncertainties in future rainfall mean that predictions of fire impacts must be conditional on climate predictions, but the link between warmth, drought and fire is universal in regions wet enough to support forest (Daniau *et al.*, 2012). Simulations suggest that fragmentation rather than climate change will be the major driver of forest fire in Amazonian landscapes over future decades but also confirm the expected synergies between the two impacts (Silvestrini *et al.*, 2011; Soares-Filho *et al.*, 2012). These synergies may be further exacerbated by the feedbacks between deforestation, smoke aerosols and reduced regional rainfall (Aragão, 2012; Tosca *et al.*, 2012).

In the longer term, fragmentation may limit the ability of the forest community to respond to change by a turnover in species composition, with the local extinction of intolerant species no longer balanced by the immigration of more tolerant species from elsewhere. Even if the gaps between fragments do not entirely block plant movements, fragmentation is expected to substantially reduce the ability of forest species to track climate change by movements along climate gradients (Corlett and Westcott, 2013). Maladapted species may persist under less favourable climatic conditions, but they are likely to store less carbon and to be increasingly susceptible to pests and pathogens. Potential interactions between forest pests and pathogens and climate change are already a major concern in temperate and boreal forests (e.g. Sturrock *et al.*, 2011; Rustad *et al.*, 2012). Predictions of species interactions are inherently difficult, but the ability of short-lived pests and pathogens to spread and adapt faster than long-lived trees provides a potential general mechanism for increasing negative impacts.

Taken together, the impacts discussed above are expected to contribute to a major reduction in the resilience of fragmented forests to climate change in comparison with continuous forest. There is a real risk that fragmentation of forests in a warming and drying climate could lead to a tipping point beyond which forest will not persist, even in a climate that could still have supported continuous forest.

Protecting and Restoring Fragments as Climate Change Mitigation and Adaptation

Forest fragments will normally store more carbon than the matrix in which they occur, in both biomass and soil, and also provide biodiversity and social co-benefits. Protecting and restoring fragments can help mitigate climate change, by reducing emissions and increasing sinks, but this does not necessarily mean that this is the best use of resources (such as carbon payments) available for these activities. If larger, better protected areas of forest are present in the vicinity, then prioritizing them will often be more efficient, unless the fragments have a higher carbon density. 'Biodiversity banking' schemes that allow many small fragments to be cleared if the impacts are 'offset' by protecting continuous forest outside the agricultural matrix would often, in theory, be a better option for biodiversity and carbon than protecting fragments (Edwards and Laurance, 2012), but there is currently little relevant experience with such schemes in the tropics.

Where only fragments are available, then the optimum use of resources will depend, in part, on the relative weight given to carbon and biodiversity. Both considerations, however, will usually favour the prioritization of large fragments and/or fragments with a large proportion of old-growth forest for protection and the use of forest restoration to buffer and link adjacent fragments. The same actions will contribute to climate change adaptation, that is, reducing the impacts of climate change. Increasing forest cover and connectivity can reduce local temperature maxima and decrease the impact of drought, facilitate movement by genes and species, maximize population sizes and thus increase the opportunity for evolutionary adaptation, and minimize non-climate stresses, thus reducing the overall risk of extinction (Corlett, 2011).

Protection of existing forest fragments is the first priority, with improved fire management an obvious opportunity in many areas (Coe *et al.*, 2013). A total ban on logging, hunting and other forms of exploitation will rarely be practical outside formally protected areas, but the adoption of reduced-impact logging and protection of key dispersal agents should be encouraged. Where increasing forest cover is an option, a wide range of restoration techniques is now available, from a passive reliance on forest succession, to active liberation of spontaneous tree regeneration, to high-density planting (Corlett, 2014). Enrichment planting of secondary forest patches is another option (Paquette *et al.*, 2009), with the planting of long-lived emergents one possible strategy for accelerating carbon accumulation (Mascaro *et al.*, 2012). 'Leaving it to nature' is cheap and easy, if fire is not a major issue, but experience across the tropics suggests that, even where embedded fragments of primary forest provide potential seed sources for a diverse native flora, secondary forests tend to be dominated initially by relatively few pioneer species (Turner *et al.*, 1997; Martinez-Garza and Howe, 2003), in some cases including invasive aliens (Ramjohn *et al.*, 2012). Although planting is expensive, it provides an opportunity to partly reverse the 'few winners and many losers' changes in fragmented landscapes, which result in the dominance of a few disturbance-adapted species at the expense of relatively intolerant forest interior species (Tabarelli *et al.*, 2012). Restoring forest cover can also restore soil carbon stocks, but more data is needed on the impacts of different reforestation strategies and techniques (Don *et al.*, 2011; Nogueira *et al.*, 2011).

It may also be worthwhile selecting species for planting that are adapted to the expected future climate of the site. This is relatively uncontroversial if planting simply favours a tolerant subset of the existing flora, but more so if it involves 'managed translocation' (also called assisted migration and assisted colonization) of populations outside their current natural range (Schwartz *et al.*, 2012). Still more controversial would be enrichment planting with non-native species from distant sources (in contrast to the 'near neighbours' that are used in managed translocation) to accelerate carbon accumulation. Indeed, if carbon is the only issue, monoculture plantations of fast-growing exotic trees may be favoured. Although there is growing evidence that spontaneous secondary forests and plantings of mixed native species can be at least as effective in carbon sequestration (Kanowski and Catterall, 2010; Sang *et al.*, 2012), single-species plantations of exotic trees have the advantages of simplicity, accumulated silvicultural experience, low planting costs and, if timber is to be harvested, known markets (Hodgman *et al.*, 2012). Conversely, a broader focus than lowest-cost carbon sequestration, with full accounting of environmental costs and benefits, is more likely to favour the preservation and restoration of mixed native forests (Lindenmayer *et al.*, 2012).

Forest protection and restoration on a scale large enough to have a significant impact on net anthropogenic carbon emissions will need funding or other incentives beyond those available at present. While REDD+, if it materializes, may seem the obvious source of support, the nature of fragments – relatively small, dispersed and often in well-populated areas – makes them particularly appropriate for local conservation initiatives, using volunteers and support from local governments and businesses. Such projects are more likely to focus on biodiversity than carbon, but carbon funding could become significant if many different fragment projects were aggregated at the regional or national level. This bottom-up approach seems likely to minimize the problems that may arise if industrial-scale carbon projects are implemented from the top down without adequate local participation (e.g. Rodrigues *et al.*, 2011; Barr and Sayer, 2012).

References

Anderson-Teixeira, K.J., Snyder, P.K., Twine, T.E., Cuadra, S., Costa, M.H. and DeLucia, E.H. (2012) Climate regulation services of natural and agricultural ecoregions of the Americas. *Nature Climate Change* 2, 177–181.

Aragão, L.O.C. (2012) The rainforest's water pump. *Nature* 489, 217–218.

Armenteras, D., González, T.M. and Retana, J. (2013) Forest fragmentation and edge influence on fire occurrence and intensity under different management types in Amazon forests. *Biological Conservation* 159, 73–79.

Barr, C. and Sayer, J. (2012) The political economy of reforestation and forest restoration in Asia-Pacific: critical issues for REDD+. *Biological Conservation* 154, 9–19.

Beck, P.S.A., Goetz, S.J., Alexander, H., Mack, M.C., Jin, Y., Randerson, J.T. and Loranty, M.M. (2011) The effects and implications of an intensifying fire regime on boreal forest in Alaska. *Global Change Biology* 17, 2853–2866.

Briant, G., Gond, V. and Laurance, S.G. (2010) Habitat fragmentation and the desiccation of forest canopies: a case study from eastern Amazonia. *Biological Conservation* 143, 2763–2769.

Broadbent, E.N., Asner, G.P., Keller, M., Knapp, D.E., Oliveira, P.J.C. and Silva, J.N.M. (2008) Forest fragmentation and edge effects from deforestation and selective logging in the Brazilian Amazon. *Biological Conservation* 141, 1745–1757.

Brodie, J.F. and Gibbs, H.K. (2009) Bushmeat hunting as climate threat. *Science* 326, 364–365.

Chabrerie, O., Jamoneau, A., Gallet-Moron, E. and Decocq, G. (2013) Maturation of forest edges is constrained by neighbouring agricultural land management. *Journal of Vegetation Science* 24, 58–69.

Coe, M.T., Marthews, T.R., Costa, M.H., Galbraith, D., Greenglass, H., Imbuzeiro, H.M.A., Levine, N.M., Malhi, Y., Moorcroft, P., Muza, M.N., Powell, T.L., Saleska, S. *et al.* (2013) Deforestation and climate feedbacks threaten the ecological integrity of south-southeastern Amazonia. *Philosophical Transactions of The Royal Society B: Biological Sciences* 368, 20120155.

Corlett, R.T. (2011) Impacts of warming on tropical lowland rainforests. *Trends in Ecology and Evolution* 2, 606–613.

Corlett, R.T. (2014) *The Ecology of Tropical East Asia*, 2nd edn. Oxford University Press, Oxford, UK.

Corlett, R.T. and Westcott, D.A. (2013) Will plant movements keep up with climate change? *Trends in Ecology and Evolution* 28, 482–488.

Daniau, A.-L., Bartlein, P.J., Harrison, S.P., Prentice, I.C., Brewer, S., Friedlingstein, P., Harrison-Prentice, T.I., Inoue, J., Marlon, J.R., Mooney, S., Power, M.J., Stevenson, J. *et al.* (2012) Predictability of biomass burning in response to climate changes. *Global Biogeochemical Cycles* 26, GB4007.

Don, A., Schumacher, J. and Freibauer, A. (2011) Impact of tropical land-use change on soil organic carbon stocks – a meta-analysis. *Global Change Biology* 17, 1658–1670.

Edwards, D.P. and Laurance, S.G.W. (2012) Green labelling, sustainability and the expansion of tropical agriculture: critical issues for certification schemes. *Biological Conservation* 151, 60–64.

Fearnside, P.M., Righi, C., Graça, P.M.L.A., Keizer, E., Cerri, C., Nogueira, E. and Barbosa, R. (2009) Biomass and greenhouse-gas emissions from land-use change in Brazil's Amazonian "arc of deforestation": the states of Mato Grosso and Rondônia. *Forest Ecology and Management* 258, 1968–1978.

Georgescu, M., Lobell, D.B., Field, C.B. and Mahalov, A. (2013) Simulated hydroclimatic impacts of projected Brazilian sugarcane expansion. *Geophysical Research Letters* 40, 1–6.

Giambelluca, T.W., Hölscher, D., Bastos, T.X., Frazão, R.R., Nullet, M.A. and Ziegler, A.D. (1997) Observations of albedo and radiation balance over post-forest land-surfaces in the eastern Amazon Basin. *Journal of Climate* 19, 919–928.

Guo, L.B. and Gifford, R.M. (2002) Soil carbon stocks and land use change: a meta analysis. *Global Change Biology* 8, 345–360.

Hallgren, W., Schlosser, C.A, Monier, E., Kicklighter, D., Sokolov, A. and Melillo, J. (2013) Climate impacts of a large-scale biofuels expansion. *Geophysical Research Letters* 40, 1624–1630.

Harrison, M.E. and Paoli, G.D. (2012) Managing the risk of biodiversity leakage from prioritizing REDD+ in the most carbon-rich forests: the case study of peat-swamp forests in Kalimantan, Indonesia. *Tropical Conservation Science* 5, 426–433.

Harrison, R.D., Tan. S., Plotkin, J.B., Slik, F., Detto, M., Brenes, T., Itoh, A. and Davies, S.J. (2013) Consequences of defaunation for a tropical tree community. *Ecology Letters* 16, 687–694.

Hiraldo, R. and Tanner, T. (2011) Forest voices: competing narratives over REDD+. *IDS Bulletin* 42, 42–51.

Hodgman, T., Munger, J., Hall, J. and Ashton, M.S. (2012) Managing afforestation and reforestation for carbon sequestration: considerations for land managers and policy makers. In: Ashton, M.S., Tyrrell, M.L., Spalding, D. and Gentry, B. (eds) *Managing Forest Carbon in a Changing Climate*. Springer, Dordrecht, the Netherlands, pp. 227–255.

Houghton, R.A. (2012) Carbon emissions and the drivers of deforestation and forest degradation in the tropics. *Current Opinion in Environmental Sustainability* 4, 597–603.

Huntingford, C., Zelazowski, P., Galbraith, D., Mercado, L.M., Sitch, S., Fisher, R., Lomas, M., Walker, A.P., Jones, C.D., Booth, B.B.B., Malhi, Y., Hemming, D. *et al.* (2013) Simulated resilience of tropical rainforests to CO_2-induced climate change. *Nature Geoscience* 6, 268–273.

IPCC (2012) *Managing the Risks of Extreme Events and Disasters to Advance Climate Change Adaptation. A Special Report of Working Groups I and II of the Intergovernmental Panel on Climate Change*. Cambridge University Press, Cambridge, UK.

Jansen, P.A., Muller-Landau, H.C. and Wright, S.J. (2010) Bushmeat hunting and climate: an indirect link. *Science* 327, 30.

Kanowski, J. and Catterall, C.P. (2010) Carbon stocks in above-ground biomass of monoculture plantations, mixed species plantations and environmental restoration plantings in north-east Australia. *Ecological Management and Restoration* 11, 19–126.

Laurance, W.F., Nascimento, H.E.M., Laurance, S.G., Andrade, A., Ribeiro, J.E.L.S., Giraldo, J.P., Lovejoy, T.E., Condit, R., Chave, J., Harms, K.E. and D'Angelo, S. (2006) Rapid decay of tree-community composition in Amazonian forest fragments. *Proceedings of the National Academy of Sciences USA* 103, 19010–19014.

Lindenmayer, D.B., Hulvey, K.B., Hobbs, R.J., Colyvan, M., Felton, A., Possingham, H., Steffen, W., Wilson, K., Youngentob, K. and Gibbons, P. (2012) Avoiding bio-perversity from carbon sequestration solutions. *Conservation Letters* 5, 28–36.

Lira, P.K., Tambosi, L.R., Ewers, R.M. and Metzger, J.P. (2012) Land-use and land-cover change in Atlantic Forest landscapes. *Forest Ecology and Management* 278, 80–89.

Martínez-Garza, C. and Howe, H.F. (2003) Restoring tropical diversity: beating the time tax on species loss. *Journal of Applied Ecology* 40, 423–429.

Mascaro, J., Asner, G.P., Dent, D.H., DeWalt, S.J. and Denslow, J.S. (2012) Scale-dependence of aboveground carbon accumulation in secondary forests of Panama: a test of the intermediate peak hypothesis. *Forest Ecology and Management* 276, 62–70.

Nogueira, L.R. Jr, Gonçalves, J.L. de M., Engel, V.L. and Parrotta, J. (2011) Soil dynamics and carbon stocks 10 years after restoration of degraded land using Atlantic Forest tree species. *Forest System* 20, 536–545.

Numata, I., Cochrane, M.A., Roberts, D.A., Soares, J.V., Souza, C.M. and Sales, M.H. (2010) Biomass collapse and carbon emissions from forest fragmentation in the Brazilian Amazon. *Journal of Geophysical Research: Biogeosciences* 115, G03027.

Paquette, A., Hawryshyn, J., Vyta Senikas, A. and Potvin, C. (2009) Enrichment planting in secondary forests: a promising clean development mechanism to increase terrestrial carbon sinks. *Ecology and Society* 14, 31.

Pütz, S., Groeneveld, J., Alves, L.F., Metzger, J.P. and Huth, A. (2011) Fragmentation drives tropical forest fragments to early successional states: a modeling study for Brazilian Atlantic forests. *Ecological Modelling* 222, 1986–1997.

Ramjohn, I.A., Murphy, P.G., Burton, T.M. and Lugo, A.E. (2012) Survival and rebound of Antillean dry forests: role of forest fragments. *Forest Ecology and Management* 284, 124–132.

Reich, P.B. (2011) Taking stock of forest carbon. *Nature Climate Change* 1, 346–347.

Ribeiro, M.C., Metzger, J.P., Martensen, A.C., Ponzoni, F.J. and Hirota, M.M. (2009) The Brazilian Atlantic Forest: how much is left, and how is the remaining forest distributed? Implications for conservation. *Biological Conservation* 142, 1141–1153.

Richter, D. deB. and Houghton, R.A. (2011) Gross CO_2 fluxes from land-use change: implications for reducing global emissions and increasing sinks. *Carbon Management* 2, 41–47.

Rodrigues, R.R., Gandolfi, S., Nave, A.G., Aronson, J., Barreto, T.E., Vidal, C.Y. and Brancalion, P.H.S. (2011) Large-scale ecological restoration of high-diversity tropical forests in SE Brazil. *Forest Ecology and Management* 261, 1605–1613.

Ross, K.A., Fox, B.J. and Fox, M.D. (2002) Changes to plant species richness in forest fragments: fragment age, disturbance and fire history may be as important as area. *Journal of Biogeography* 29, 749–765.

Rustad, L., Campbell, J., Dukes, J.S., Huntington, T., Lambert, K.F., Mohan, J. and Rodenhouse, N. (2012) *Changing Climate, Changing Forests: the Impacts of Climate Change on Forests of the Northeastern United States and Eastern Canada.* General Technical Report NRS-99. US Department of Agriculture, Forest Service, Pennsylvania.

Sang, P.M., Lamb, D., Bonner, M. and Schmidt, S. (2012) Carbon sequestration and soil fertility of tropical tree plantations and secondary forest established on degraded land. *Plant and Soil* 362, 187–200.

Schnitzer, S.A. and Bongers, F. (2011) Increasing liana abundance and biomass in tropical forests: emerging patterns and putative mechanisms. *Ecology Letters* 14, 397–406.

Schwartz, M.W., Hellmann, J.J., McLachlan, J.S., Sax, D.F., Borevitz, J.O., Brennan, J., Camacho, A.E., Ceballos, G., Clark, J.R., Doremus, H., Early, R., Etterson, J.R. *et al.* (2012) Managed relocation: integrating the scientific, regulatory, and ethical challenges. *BioScience* 62, 732–743.

Senici, D., Chen, H.Y.H., Bergeron, Y. and Cyr, D. (2010) Spatiotemporal variations of fire frequency in central boreal forest. *Ecosystems* 13, 1227–1238.

Shuman, J.K., Shugart, H.H. and O'Halloran, T.L. (2011) Sensitivity of Siberian larch forests to climate change. *Global Change Biology* 17, 2370–2384.

Silvério, D.V., Brando, P.M., Balch, J.K., Putz, F.E., Nepstad, D.C., Oliveira-Santos, C. and Bustamante, M.M.C. (2013) Testing the Amazon savannization hypothesis: fire effects on invasion of a neotropical forest by native cerrado and exotic pasture grasses. *Philosophical Transactions of The Royal Society B: Biological Sciences* 368, 20120427.

Silvestrini, R.A., Soares-Filho, B.S., Nepstad, D., Coe, M., Rodrigues, H. and Assunção, R. (2011) Simulating fire regimes in the Amazon in response to climate change and deforestation. *Ecological Applications* 22, 1573–1590.

Smith, P., Davies, C.A., Ogle, S., Zanchi, G., Bellarby, J., Bird, N., Boddey, R.M., McNamara, N.P., Powlson, D., Cowie, A., van Noordwijk, M., Davis, S.C. *et al.* (2012) Towards an integrated global framework to assess the impacts of land use and management change on soil carbon: current capability and future vision. *Global Change Biology* 18, 2089–2101.

Soares-Filho, B., Silvestrini, R., Nepstad, D., Brando, P., Rodrigues, H., Alencar, A., Coe, M., Locks, C., Lima, L., Hissa, L. and Stickler, C. (2012) Forest fragmentation, climate change and understory fire regimes on the Amazonian landscapes of the Xingu headwaters. *Landscape Ecology* 27, 585–598.

Spracklen, D.V., Arnold, S.R. and Taylor, C.M. (2012) Observations of increased tropical rainfall preceded by air passage over forest. *Nature* 489, 282–285.

Stocker, B.D., Roth, R., Joos, F., Spahni, R., Steinacher, M., Zaehle, S., Bouwman, L. and Prentice, I.C. (2013) Multiple greenhouse-gas feedbacks from the land biosphere under future climate change scenarios. *Nature Climate Change* 3, 666–672.

Sturrock, R.N., Frankel, S.J., Brown, A.V., Hennon, P.E., Kliejunas, J.T., Lewis, K.J., Worrall, J. J. and Woods, A.J. (2011) Climate change and forest diseases. *Plant Pathology* 60, 133–149.

Tabarelli, M., Peres, C.A. and Melo, F.P.L. (2012) The 'few winners and many losers' paradigm revisited: emerging prospects for tropical forest biodiversity. *Biological Conservation* 155, 136–140.

Tosca, M.G., Randerson, J.T., Zender, C.S., Flanner, M.G. and Rasch, P.J. (2010) Do biomass burning aerosols intensify drought in equatorial Asia during El Niño? *Atmospheric Chemistry and Physics* 10, 3515–3528.

Tosca, M.G., Randerson, J.T. and Zender, C.S. (2012) Global impact of contemporary smoke aerosols from landscape fires on climate and the Hadley circulation. *Atmospheric Chemistry and Physics Discussion* 12, 28069–28108.

Turner, I.M., Wong, Y.K., Chew, P.T. and bin Ibrahim, A. (1997) Tree species richness in primary and old secondary tropical forest in Singapore. *Biodiversity and Conservation* 6, 237–543.

Wicklein, H.F., Carter, M.E., Christopher, D. and Smith, B.H. (2012) Edge effects on sapling characteristics and microclimate in a small temperate deciduous forest fragment. *Natural Areas Journal* 32, 110–116.

7 Forest Islands in an Agricultural Sea

Alexandra-Maria Klein,[1,2*] Virginie Boreux,[1,2] Jürgen Bauhus,[2] M. Jahi Chappell,[3] Joern Fischer[1] and Stacy M. Philpott[4]
[1]*Leuphana University of Lüneburg, Germany;* [2]*University of Freiburg, Germany;* [3]*Washington State University, Vancouver, USA;* [4]*University of California, Santa Cruz, USA*

Abstract

Under the current scenario of continuing human population growth, achieving high agricultural yield while conserving forest biodiversity is challenging scientists and policy makers alike. There is ongoing debate as to whether biodiversity should be integrated on the same land (land sharing) or separated from agriculture (land sparing). Here, we present examples of land-sharing practices (agroforestry systems and multipurpose forests) and demonstrate that forest biodiversity conservation does not necessarily compromise crop yield. There is no simple trade-off between biodiversity and yield, and the complexity of crop yield–biodiversity relationships has not yet been fully investigated, at either the habitat or the landscape scale. We argue that land-sparing management has to be considered at different spatial scales. Agriculture in large-scale land sparing (intensive agriculture separated from forest remnants) generally does not benefit much from ecosystem services mediated by forest biodiversity, but small forest remnants can enhance biodiversity in large agricultural and forestry plantations. Small-scale land sparing (intensive agriculture connected to forest remnants) promotes beneficial organisms and their associated ecosystem services. In such landscapes, wild beneficial organisms, e.g. a diversity of pollinators, can enhance and stabilize crop pollination in intensively managed fields due to spillover from forests into crop fields. Functional connectivity of forest remnants (single trees, treelines, hedgerows and forest patches) in high-input agricultural landscapes can enhance biodiversity and ecosystem services, but future research needs to consider connectivity at various land-sparing scales to evaluate their conservation effectiveness. Furthermore, landscape developments minimizing possible adverse consequences of forest conservation on crop production are currently often overlooked but need to be considered in future research and landscape planning. We argue that both land sharing and land sparing can promote biodiversity without compromising high yields, and that a combination of management strategies at different spatial scales, including the maintenance of forest connectivity, may most effectively safeguard both biodiversity and livelihood security.

Conciliating Agricultural Production and Forest Conservation

The human population has increased from 3 to 7 billion people in the past 50 years, while agriculture now covers almost 40% of the total earth's ice-free land surface (Ellis *et al.*, 2010; see Chapter 3, this volume). With the global population expected to increase to 9 billion by 2050, the pressure on natural habitats, in particular forests, is growing (Garibaldi *et al.*, 2011a; Roberts, 2011). Deforestation reached

*alexandra.klein@nature.uni-freiburg.de

13 million ha per annum in 2010, much of it attributed to the establishment of agricultural areas (FAO, 2010) and taking place in tropical and subtropical ecosystems where high growth rates in human populations occur today and are predicted for the near future (Bloom, 2011; see Chapter 3, this volume).

Considering the benefits forests provide to humans worldwide, through the provision of ecosystem services (e.g. carbon sequestration, climate regulation, crop pollination) and traditional forest resources (e.g. timber, fibre and edible plant materials like fruits), we urgently need to address biodiversity conservation in combination with human livelihoods (see Chapter 9, this volume).

A societal issue

Decisions on agricultural production, biodiversity and human livelihood have to be considered in their social context, where values weigh as much as scientific principles or results (Jasanoff, 2004; Layzer, 2011). Trade-offs related to forest conservation and food production, although based on scientific arguments, rely heavily on the values and personal judgements of researchers (see Shrader-Frechette, 1995a, on ecological decision making; Shrader-Frechette, 1995b, on risk assessment; Metzger *et al.*, 2010, on scenario-modelling). For instance, in an agroecological approach, improving human livelihoods and biodiversity conservation is addressed in terms of trade-offs in food production, i.e. our aim is to produce a given amount of food as efficiently as possible, while conserving biodiversity. This perspective overlooks decades of research showing that food production has only limited impacts on human livelihood. Indeed, malnutrition today is affecting about 850 million people worldwide even though the overall global production is sufficient (Smith *et al.*, 2000; Smil, 2001; Chappell and LaValle, 2011). Lundqvist *et al.* (2008) indicated that an average of 4600 kcal per person per day is available for consumption (2800 kcal per person per day after wastage and conversion to livestock). Even the lower of the two estimates corresponds to the higher end of the recommended daily intake (recommended daily intakes range from 1800 to 2800 kcal per person per day; FAO, 2013).

However, food production and human livelihood are still generally a local and distributional issue. Even with enough food produced at the global scale, many developing countries need to shift their agricultural practices to produce enough food for local markets in a sustainable way. This needs to integrate socio-economic conditions into the decision process to achieve long-term biodiversity conservation and livelihood security.

The concept of land sharing and land sparing

In the last decade, this has been approached from two contrasting and often opposed perspectives, namely land sharing versus land sparing (Green *et al.*, 2005).

Land sharing promotes the conservation of biodiversity on the same land used for agriculture or forest production systems, requiring lower-intensity practices but allowing agricultural and forest products to grow on large areas of land. This approach is associated with biodiversity-friendly farming practices such as organic management or multiple cropping, and is usually associated with lower production per unit area (Seufert *et al.*, 2012). As less area is specifically allocated for conservation purposes, this approach can fail to conserve specialist species associated with undisturbed forests. However, landscapes managed under land sharing often include interspersed forest patches acting as biodiversity repositories (Fischer *et al.*, 2008; Wright *et al.*, 2012).

Land sparing focuses on biodiversity conservation in designated reserve areas requiring agricultural or forest product development to be focused in restricted but highly intensified areas. This is typically, although not necessarily, achieved through higher inputs of agro-chemicals and lower crop diversity. High production outcomes on any given piece of land make it, in theory, economically feasible to set aside areas for biodiversity conservation. Landscapes managed under the land-sparing approach/concept are dominated by large areas of intensive agriculture or plantation forests and protected areas such as native forests. Such forests are, however, typically not well connected with one another, nor with the farming or plantation system.

Land sparing and land sharing can be considered end points of a continuum of scales at which biodiversity conservation and agricultural and forest production systems can be integrated. Whether either of these two end points or instead the whole continuum should be promoted to minimize trade-offs between human livelihoods and biodiversity conservation has been discussed extensively (e.g. Fischer *et al.*, 2011; Phalan *et al.*, 2011). Several studies argue for the importance of land sparing in biodiversity conservation (Waggoner, 1996; Egan and Mortensen, 2012), particularly for species sensitive to even relatively low levels of disturbance (Green *et al.*, 2005) or with habitat restrictions in terms of both range and habitat requirements (Aratrakorn *et al.*, 2006). In the forestry context it has been argued that the intensification of production in plantations on a small proportion of land offers the opportunity to set aside larger areas of native forests for conservation (Bauhus *et al.*, 2010). Others have discussed the benefits of land sharing (Dorrough *et al.*, 2007; Perfecto and Vandermeer, 2010; Wright *et al.*, 2012) and the positive interaction between agricultural production and ecosystem services delivered by on-site biodiversity (Whittingham, 2011).

Both approaches have advantages and drawbacks (see Matson and Vitousek, 2006; Ewers *et al.*, 2009; Fischer *et al.*, 2011; Gutiérrez-Vélez *et al.*, 2011), and to a large extent, their relative merits depend on both the biophysical characteristics and the socio-economic context of any given landscape (Fischer *et al.*, 2008, 2011). In the following sections, we will discuss the potential benefits and issues associated with land sharing and land sparing using case studies from tropical agroforestry systems. Focusing first on land sharing, we will show that the trade-off between production and biodiversity conservation is invalidated in some cases. We will then show that the land-sparing approach can be beneficial to agricultural production, particularly at smaller scales. In this context, we will define the concept of connectivity between natural habitats and discuss the use of corridors and stepping stones to support both biodiversity conservation and agricultural production in the land-sparing approach.

Land Sharing: Agroforestry Systems and Multipurpose Forestry Plantations

The example of traditional agroforestry systems

Agroforestry systems are defined as crops and forest trees growing in the same field. Certain multistrata agroforestry systems, including coffee and cacao, provide important habitats for biodiversity in fragmented tropical landscapes (Philpott *et al.*, 2007; Tscharntke *et al.*, 2011). Such 'cash' crops are not providing direct sustenance to the planters but contribute greatly to their livelihoods, as both coffee and cacao are high-priced luxury goods. Moreover, fruit trees as well as medicinal plants are also grown in the agroforestry systems, contributing to the socio-economic stability and food security of the family (Jaffee, 2007; Méndez *et al.*, 2010). Coffee and cacao were traditionally cultivated under a diverse and dense canopy of native forest trees, making these agroforestry systems the most biodiverse across agroecosystems (Perfecto *et al.*, 1996; Philpott *et al.*, 2008). However, ongoing intensification of management in these systems (e.g. shade tree removal, use of agrochemicals) has negatively affected the biodiversity of plants, arthropods, birds and mammals (Moguel and Toledo, 1999; Perfecto *et al.*, 2003; Philpott *et al.*, 2008; Tscharntke *et al.*, 2011). Much attention has been placed on the value of traditional agroforestry systems for biodiversity conservation, but a more complete understanding of the interactions between biodiversity conservation, shade cover, yield and overall livelihood value of the agroforestry systems is required to promote the use of coffee and cacao agroforestry systems for conservation.

Shade, yield and biodiversity in traditional agroforestry systems

Farmers often convert their traditional, shaded agroforestry systems to intensive sun farms because reducing or eliminating shade is presumed to increase yield. However, empirical evidence on interactions between yield and shade density is far from consistent. According to some studies, moderate shade levels do not negatively affect coffee or cacao yield (Wood and Lass, 2001; Perfecto *et al.*, 2005) and in some cases promote production compared with higher or lower canopy covers (Soto-Pinto *et al.*, 2000), while others document higher cacao yield with the elimination of canopy cover, especially for young plants (Steffan-Dewenter *et al.*, 2007; Tscharntke *et al.*, 2011). The results probably depend on study region, crop variety, crop plant age, shade tree species and other variables. Mounting evidence describes relationships between shade management and biodiversity, but there is little discussion of how biodiversity directly relates to yield. Because reducing shade results in biodiversity loss (Philpott *et al.*, 2008), conventional knowledge on agroforestry management assumes a strong trade-off between biodiversity and yield (Moguel and Toledo, 1999; Perfecto *et al.*, 2005).

Cacao agroforestry systems

The relationship between biodiversity and yield in cacao agroforestry systems has been studied in Indonesia and Cameroon, two major cacao-producing countries, as well as in Ecuador. In Sulawesi, Indonesia, Steffan-Dewenter *et al.* (2007) investigated changes in yield, net revenue, biodiversity (of herbs, bryophytes, trees, ants, beetles, wasps and bees) and a range of ecosystem services (biomass accumulation, fine root biomass, carbon sequestration, litter decomposition rate, parasitism rate) in both high-shade (80%) and moderate-shade (40%) cacao agroforests. Reducing the canopy minimally affected biodiversity, maintained a relatively high provision of ecosystem services and increased farmer income twofold. However, completely eliminating shade increased farmer income by a further 40%, while strongly negatively affecting biodiversity and ecosystem services. Thus, some shade modifications lead to win–win situations where both biodiversity and yield are maintained, but there may be thresholds beyond which this is no longer true.

In the same area, Clough *et al.* (2011) examined the biodiversity of nine taxa (trees, herbs, endophytic fungi, butterflies, ants, spiders, birds, small mammals and reptiles) and economic information extracted from farmer surveys. Study plots ranged in shade management from low-intensity agroforestry systems to intensively managed tree-crop plantations, and in distance to forest. Herb biodiversity was negatively correlated with yield, while diversity of the other eight taxa did not vary with yield (Fig. 7.1). Species richness strongly responded to distance to forest and properties of the shade canopy. In contrast, yield was negatively correlated with percentage shade cover, but cacao disease was a more important predictor of yield loss. Finally, Waldron *et al.* (2012) examined relationships between shade cover, yield, net incomes and bird biodiversity in the Ecuadorian Chocó. Cacao yield (per tree and per hectare) increased with increasing canopy cover up to 40% canopy cover. Likewise, bird diversity increased with shade cover, with a tipping point just slightly higher than that of the cacao yield. Net income increased linearly with increasing shade cover. They attributed this linear relationship to increased ecosystem services (e.g. weed control) in shaded farms that allowed cacao farmers the time to take off-farm jobs and supplement their total incomes.

Coffee agroforestry systems

For coffee, nearly all studies examining the relationship between yield and biodiversity are from Mexico or Costa Rica. Romero-Alvarado *et al.* (2002) compared yield in farms in Chiapas, Mexico, that varied in tree richness and other shade attributes. They found no decline in coffee yield with decreasing shade tree diversity but documented higher weed pressure in farms with low tree diversity. Perfecto *et al.* (2005) combined data from two separate studies from Chiapas to examine biodiversity–yield relationships. In the first,

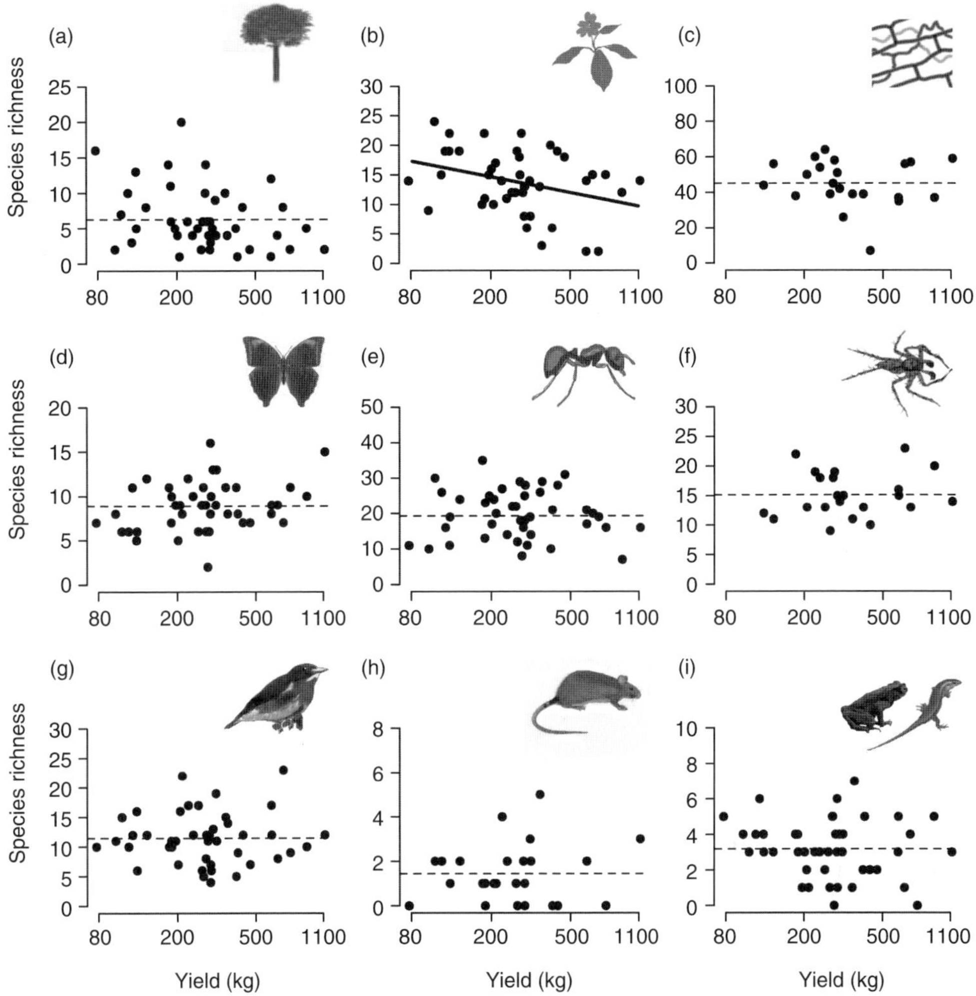

Fig. 7.1. Non-crop biodiversity in smallholder cacao agroforests does not decrease with yield in trees (a), endophytic fungi (c), butterflies (d), ants (e), spiders (f), birds (g), rats (h) and amphibians (i), but decreases significantly in herbs (b). Broken lines are intercept-only linear models. Morphospecies richness is indicated for endophytic fungi, ants and spiders. For tree, butterfly, ant, bird, amphibian, and reptile data, $n = 43$; for endophytic fungi, spider, and rat data, $n = 22$. (From Clough *et al.*, 2011.)

Soto-Pinto *et al.* (2000) documented maximum coffee yield at 48% canopy cover, with declines at lower and higher canopy cover. In the second, Perfecto *et al.* (2003) documented declines in butterfly richness with reductions in canopy cover but lesser effects on ants. Combining biodiversity and yield data from the two studies showed that the relationship between biodiversity loss and yield loss was not linear and strongly depended on whether butterflies or ants were the focal taxon. For example, maintaining 75% of the coffee yield would maintain 80% of forest ant species but only 23% of the butterfly species. Finally, Gordon *et al.* (2007) found higher richness of birds in diverse, traditional coffee farms in Veracruz, Mexico, and declines in bird diversity with increasing intensification indicated by negative correlations with shade cover loss, epiphyte loss and reductions in canopy height. However, farm profitability (a measure of yield combined with coffee price) did not decline with increasing shade. Furthermore, they documented a quadratic relationship between yield and bird richness, with highest yield at high and low levels of bird richness.

Wood and fibre in multipurpose forestry plantations and agroforestry systems

Land sharing also comprises forested areas producing forest products (e.g. wood and fibre) together with old-growing forest trees.

Around 85% of the global forest area is neither formally protected nor used for intensively managed plantations, and 55% of the global forest area is and will continue to be, managed for multiple environmental, economic and cultural values (FAO, 2010; Lindenmayer *et al.*, 2012). These multipurpose forests also represent the matrix required to support the migration of species and flow of genes to support meta-populations across reserves and landscapes (Lindenmayer and Franklin, 2002) and diversify income sources and increase resilience and stability (Knoke *et al.*, 2005), a particularly crucial aspect in developing countries. Thus, a threefold approach consisting of the three different elements – strict reserves, intensively managed plantations and multipurpose forests – has been proposed for the sustainable management of forested landscapes (Hunter and Calhoun, 1996). It has been recognized that the management of multipurpose forests needs to complement conservation goals of strict reserves besides producing wood, fibre and other functions. Retention forestry has been postulated as a concept to combine protected forest remnants and structural elements such as single or grouped habitat trees or dead wood within managed forests to achieve the complementary role of multipurpose forests to nature reserves at different spatial scales (Gustafsson *et al.*, 2012). While there is a large body of scientific evidence in support of the retention approach (Gustafsson *et al.*, 2012; Lindenmayer *et al.*, 2012), the interactions between the different forms of management (reserves, plantations and multipurpose forests) are not yet well understood.

Plantation products can also be integrated in agroforestry systems. For example, Larbi *et al.* (2009) identified shrubs for dryland agroforestry systems in West Asia with potential to produce a high quantity of fodder and fuel wood, besides stabilizing sand dunes and sequestering carbon. Sisi *et al.* (2012) demonstrated that lucerne intercropping in a cottonwood forest (*Populus nigra* L.) increased the growth rate of fibre trees and the diameter and wall thickness of the fibre. As with some coffee and cacao varieties, abaca (*Musa textilis* L. Nee) achieves optimum fibre yields in 50% shade and is therefore a suitable fibre crop for multipurpose forests (Bande *et al.*, 2013).

Land Sparing: Forest Remnants in Agricultural Landscapes

Conceptual issues with land sparing

An underlying assumption (also referred to as the 'Borlaug hypothesis') of the land-sparing approach is that once sufficient food is produced, the rate of conversion of natural habitats to agriculture will slow or halt. Yet this is in contradiction to the unlimited growth theory that most political and economic institutions abide by (Pelletier, 2010) and has been consistently rejected as a general mechanism (Angelsen and Kaimowitz, 2001; Dorrough *et al.*, 2007; Ewers *et al.*, 2009; Rudel *et al.*, 2009; Perfecto and Vandermeer, 2010; Gutiérrez-Vélez *et al.*, 2011). Even supporters of the land-sparing approach have called it a weak process occurring under a limited set of circumstances (Ewers *et al.*, 2009), while in practice, increasing yield often increases the rate of deforestation (Angelsen and Kaimowitz, 1999, 2001; Lee *et al.*, 2001). Technological changes that increase yield or lower input costs usually influence landowners to expand on a system that has become more economically efficient (i.e. the 'Jevons Paradox'; see Foster, 2009). To efficiently implement land sparing and avoid such backfiring, it appears crucial to discuss the principles underlying the concept with stakeholders, in particular producers, and include the different stakeholders in the landscape management plan.

Whether the intensification of production systems on part of the land permits the sparing of native ecosystems for conservation purposes on another part depends very much on the legal, institutional and political context. The focus of timber production on plantations of exotic tree species and the protection of all native forests on public land in New Zealand may be viewed as an example of a successful land-sparing approach, although it does not resolve all problems of biodiversity conservation.

Forest islands as biodiversity repositories

Forest islands or remnants, although smaller than the original forests they are derived from, harbour important biodiversity of both fauna and flora (see Chapters 4 and 5, this volume). By providing various food and habitat resources throughout the year, these forests act as refuges for species unable to survive in intensive agricultural landscapes (Castelletta *et al.*, 2000; Koh and Sodhi, 2004; Koh and Wilcove, 2008; Ferger *et al.*, 2013). While officially protected forests may provide the highest diversity of fauna and flora, Bhagwat *et al.* (2005) found that other forests, such as traditional sacred groves in India, substantially contribute to biodiversity conservation, sometimes to a greater extent than protected areas. Although large forests are likely to shelter more species, smaller forest remnants, with a limited number of resources, are also important and should not be overlooked in their role as biodiversity repositories (Pardini *et al.*, 2005; Benedick *et al.*, 2006).

Spillover of organisms from forests to agriculture: ecosystem services to agriculture

Remnants of natural and semi-natural habitats in agricultural landscapes provide ecosystem services to crops. Pollination services, for example, are strongly correlated with forest remnants in the surroundings. Diversity and abundance of flower visitors in coffee (both *Coffea arabica* and *Coffea canephora*) are higher closer to large forest areas or forest remnants, leading to an increase in coffee production (Klein *et al.*, 2003; Ricketts, 2004). In addition, larger forests dampen the decline in pollinator abundance with increasing distances (Boreux *et al.*, 2012). This effect of forest remnants and other natural habitats has also been observed in temperate and Mediterranean climates, in both annual and perennial crops. In watermelon, for example, the diversity and abundance of native pollinators was greatest closest to remnants of chaparral shrubland and forest (Kremen *et al.*, 2002), while in apple, wild bee diversity and abundance increased with both distance to and size of the deciduous and mixed conifer/deciduous forests (Watson *et al.*, 2011). These results may stem from what pollinators use as nesting sites (dead and living trees) and nesting materials (leaves, twigs, tree resin), as well as year-long foraging resources in forests (e.g. stingless bees; see Eltz *et al.*, 2003). The effect of deciduous forest remnants on pollinators varies depending on the type of pollinators. Forest cover within a 3 km radius positively influenced the abundance of native honeybees in buckwheat fields, while non-honeybee visitors in the same crop were affected by the area of forest cover (as well as that of grassland) within a 100 m radius (Taki *et al.*, 2010). This shows that spillover effects from forest remnants vary with species, a critical aspect that needs to be taken into account when considering ecosystem services in land-sparing landscapes (see Chapter 4, this volume).

Forest remnants also improve the stability of pollination services spatially and temporally. For watermelon (Kremen *et al.*, 2004), visitation by native bees was more stable through time in fields closer to chaparral remnants. For coffee, spatial and temporal variability of flower-visitor diversity increased with increasing distance to the primary rainforest, resulting in lower and less predictable fruit set (Klein, 2009). A recent meta-analysis based on 29 studies (Garibaldi *et al.*, 2011b) further confirmed that the average, temporal and spatial variability of flower-visitor richness increased with increasing distance from natural habitats, regardless of climatic conditions (Fig. 7.2). By supporting

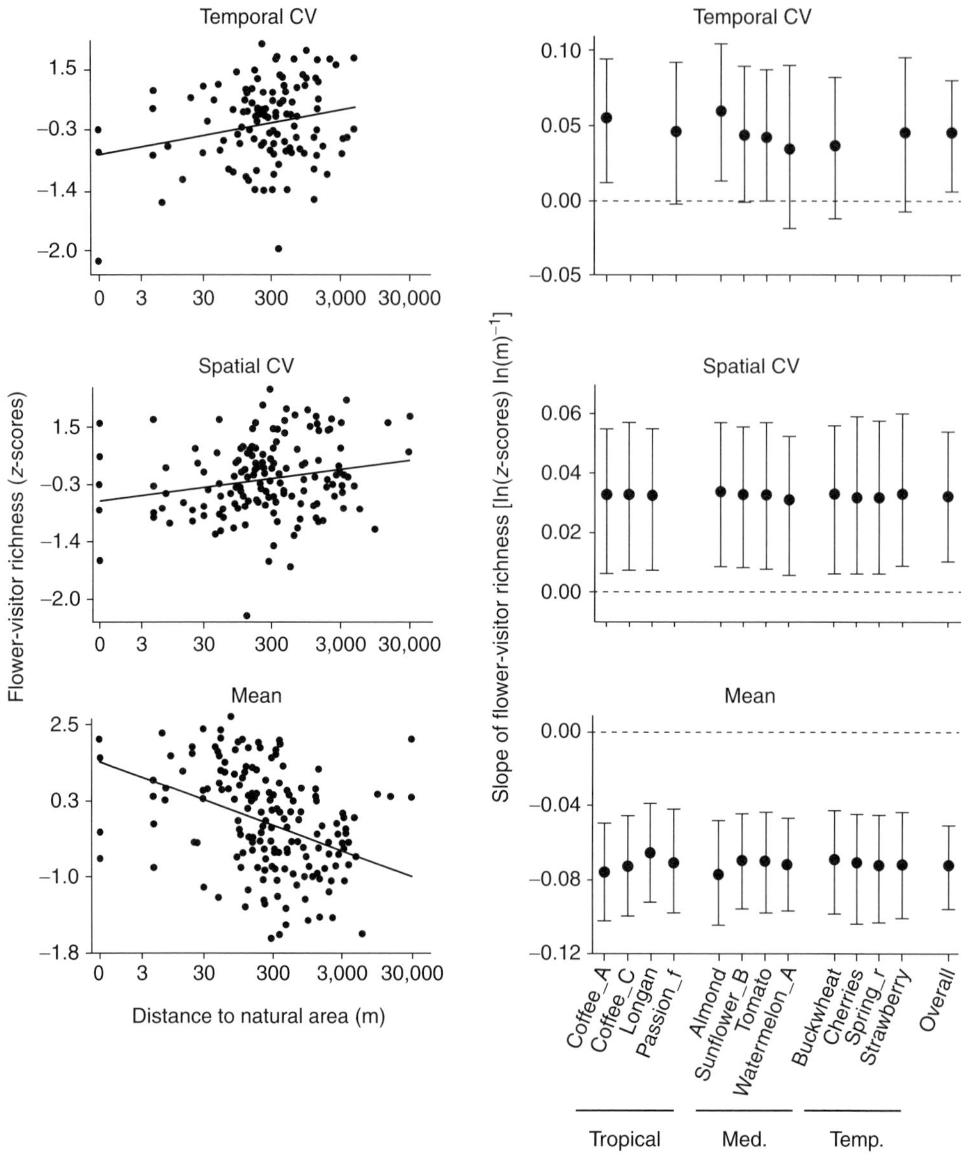

Fig. 7.2. Mean and coefficient of variation (CV) of flower-visitor richness (all insects except the European honeybee, *Apis mellifera*) in croplands in relation to distance from natural or semi-natural areas. Left: each point is a site within a study (all studies are shown) and the line presents the overall linear model estimation. Data from different studies were standardized by *z*-scores prior to analysis. Right: slopes (mean and 90% credible interval as circles and lines, respectively) for each study and overall mean. The *x*-axis of the right-hand panel shows studies from tropical and subtropical (Tropical), Mediterranean (Med.) and temperate (Temp.) regions. (From Garibaldi *et al.*, 2011b.)

larger and more diverse insect populations, forests remnants help buffer year-to-year variations in pollinator populations, which can vary greatly (Kremen *et al.*, 2004; see also Chapter 8, this volume).

Natural pest control is another critical service that may be related to proximity to forest remnants (see Chapter 8, this volume). This ecosystem service is provided by a range of organisms (from birds to arthropods) that can dramatically reduce pest populations, thereby reducing the costs of chemical pest control and reducing crop damage. Forest remnants, by providing reproduction sites and food resources, play an important role in sustaining predator and parasitoid populations in agricultural landscapes. For example, aphids in soybean fields are better controlled by their natural enemies when the crop is located in a landscape with higher rates of non-crop habitat, including forests and grasslands (Gardiner *et al.*, 2009). Similarly, predatory wasp abundance in maize fields and natural parasitism rates in coffee/cacao agroforestry systems decrease with increasing distance to the crop fields (Klein *et al.*, 2006; Santana Sousa *et al.*, 2011).

Issues of scale: habitat connectivity in agricultural landscapes

The benefits that forest remnants provide to crops are correlated with the distance and size of these forests, and therefore depend on the landscape matrix. For example, a single forest isolated in a matrix of continuous agricultural crops (referred to hereafter as 'large-scale land sparing') will most likely have a lower impact in terms of ecosystem service provision to crops than a series of smaller forests spread across the agricultural landscape, referred to hereafter as 'small-scale land sharing' (Fig. 7.3).

To compensate for the reduction in ecosystem services due to isolation from natural habitats, smaller natural features could be used to increase connectivity between forest remnants, hence facilitate the movements of beneficial species across the agricultural matrix (Fig. 7.3). For instance, lines of trees and hedgerows in agricultural landscapes establish physical linkages between isolated forest remnants, hence creating a 'structural connectivity' in the landscape. These structures have been shown to improve the movements of particular species or functional groups between distant habitats across the landscape (Gilbert *et al.*, 1998; Levey *et al.*, 2005), thus improving the 'functional connectivity', sometimes also termed 'habitat connectivity' (Uezu *et al.*, 2005).

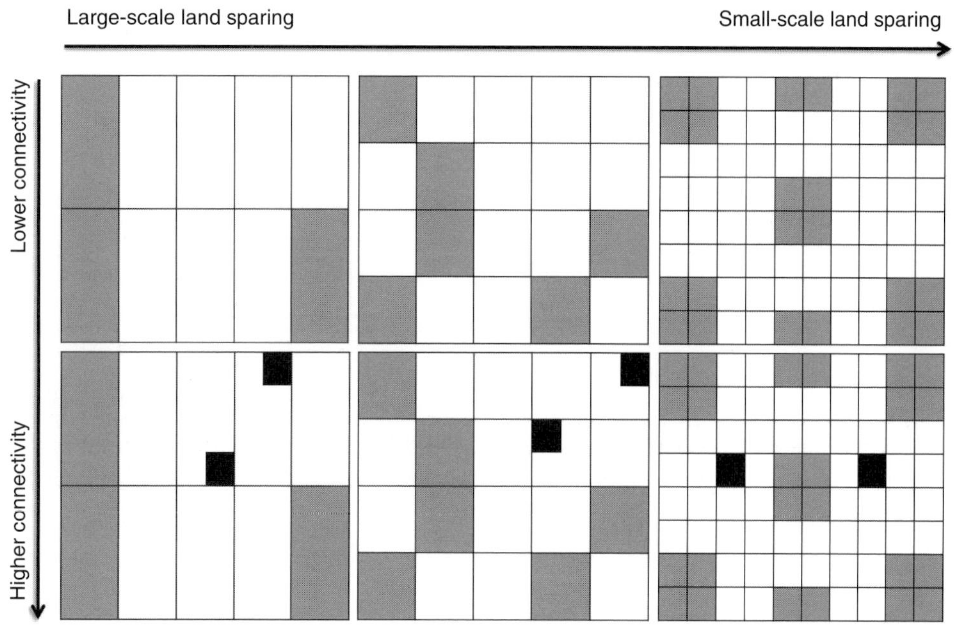

Fig. 7.3. Different land-sparing scales and associated landscape connectivity between forest patches. Agricultural lands are in white, forests in grey and connecting elements in black. The different landscapes cover the same surface area. Small-scale land sparing reduces distances between forest remnants, and enhances connectivity, spillover of organisms and ecosystem services delivery to agriculture.

Corridors have been debated by conservationists as potential management tools for over two decades (Noss, 1987; Simberloff *et al.*, 1992; Beier and Noss, 1998). Proponents of connectivity have argued that most ecosystems used to be unfragmented, and hence reconnecting natural habitats within a given landscape (Noss and Beier, 2000), or even at a continental scale (Mackey *et al.*, 2008), is likely to benefit many species. Critics have argued that corridors must not be seen as a panacea for biodiversity conservation (Simberloff *et al.*, 1992), as their establishment can be costly, and in some circumstances, scarce conservation resources may be better spent in different ways. Corridors may not be used by all species, thereby potentially improving habitat connectivity for only a small subset of species. Whether a target species will actually use a given corridor depends on a wide range of factors, including the biophysical characteristics of the corridor and the behaviour of the species (Lindenmayer, 1994). In some agricultural landscapes, riparian vegetation can provide natural corridors that can be complemented by living fences, which are common features in tropical agricultural landscapes (Harvey et al., 2006).

Stepping stones, for example individual trees, have been studied less extensively than corridors. By their very nature, stepping stones are not physically connected to a source patch, and are primarily useful for species able to cross at least small stretches of agricultural land. In many agricultural landscapes, and especially in areas used for livestock grazing, isolated remnant trees can provide important stepping stones (Fig. 7.4). For example, a radio-tracking study in the Coto Brus province of Costa Rica found that the silver-throated tanager (*Tangara icterocephala* Bonaparte) was detected in remnant trees approximately one-third of the time, even though such trees accounted for less than 5% of land cover (Sekercioglu *et al.*, 2007). Similarly, Guevara and Laborde (1993) found that daily foraging movements of frugivorous birds in a livestock grazing landscape in Mexico closely followed the distribution of remnant trees. A 'disproportionate' effect of scattered trees is increasingly being documented around the world. Manning *et al.* (2006) have termed scattered trees 'keystone structures', precisely because a single tree has a proportionally large ecological effect compared with its small footprint in the landscape (for examples

Fig. 7.4. A scattered forest tree (a) and a landscape with scattered naturally growing Macauba palm trees (*Acrocomia aculeata* (Jacq.) Lodd. Ex Mart.) on pastures and forest remnants (b) near Carmo do Paranaíba, Brazil. Scattered trees provide connectivity between forest patches in agricultural-dominated landscapes for the movement of wildlife and to enhance agroecosystem services. Photo credits: Mirco Plath, Leuphana University, Germany.

of a similar phenomenon in natural savannahs, see Tews *et al.*, 2004). While scattered trees may interfere with major infrastructure such as harvesting machinery used in large-scale agricultural industrialization, they can readily be incorporated into small-scale farming and grazing, for example in small-scale land-sparing landscapes (Figs 7.3 and 7.4), and thus improve the benefits crops get from ecosystem services delivered by species using these stepping stones. In contrast, large-scale land-sparing landscapes will benefit little, if at all, from ecosystem services mediated by the cross-habitat spillover of beneficial organisms.

Forest Conservation and Agroecosystem Services and Dis-services

Both land-sharing and land-sparing practices aim to promote biodiversity, and if such biodiversity is well connected to agricultural fields, natural habitats such as forests can improve yield due to the positive effects of biodiversity on ecosystem functions and services (Naeem *et al.*, 2009; see also Chapter 8, this volume). As mentioned previously, in agroforestry systems, pollination services benefit from bee diversity and pest control services are enhanced where predator diversity is higher (Van Bael *et al.*, 2008; Philpott *et al.*, 2009). Other services, such as nutrient cycling and decomposition, also increase with higher biodiversity, such as when a more diverse mix of plants provides leaf litter (Hooper and Vitousek, 1998; Tscharntke *et al.*, 2011). In addition, biodiversity can mediate ecosystem services provided by organisms as species are using different spatial, temporal and conditional niches leading to complementarity effects in the provisioning of ecosystem services (Naeem *et al.*, 2009; Blüthgen and Klein 2011). For instance, in Californian almond landscapes, only orchards with a high proportion of chaparral (>10% within 1 km) were visited by wild bees and maintained high pollinator diversity. Furthermore, different pollinator species visited different parts of an almond tree (upper, lower, outer and inner parts); thus, pollinator diversity was critical for pollination services (Brittain *et al.*, 2012). Additionally, the managed European honeybee, *Apis mellifera* L., did not forage in high winds, while wild pollinators like the ground-nesting solitary bee, *Andrena cerasifolii* Cockerell, visited almond flowers in high winds (Brittain *et al.*, 2012). In almond orchards with high pollinator diversity, the European honeybee provided better pollination services than in orchards with only European honeybees (Brittain *et al.*, 2013). Similar results are described in other crop systems and the authors speculate that species-specific interactions are driving these patterns, but further research is needed to justify this assumption (Klein *et al.*, 2009; Brittain *et al.*, 2013).

Given that biodiversity can increase and stabilize crop yield through ecosystem services (Garibaldi *et al.*, 2013) and that richness is higher close to forests and other natural habitats, agriculture and biodiversity should not be separated. The value of biodiversity to mediate ecosystem services needs to be considered in land-sharing and land-sparing biodiversity–production trade-off scenarios.

Very few studies have taken into account the possible trade-offs incurred by maintaining natural ecosystems for biodiversity-mediated ecosystem services. For example, although protecting forest remnants within agricultural landscapes maintains wildlife-mediated ecosystem services that in turn benefit crop production, it can also facilitate unwanted processes that negatively affect the production cycle. Such dis-services include seed predation and herbivory but also competition for the same resources such as water or pollinators (Zhang *et al.*, 2007). In addition, corridors could inadvertently facilitate the spread of fire, pathogens, weeds or other introduced species (Simberloff *et al.*, 1992). Thus, it is crucial to document whether conserving and restoring forest remnants (including corridors) in agricultural landscapes results in a net benefit for crops, taking into account pollination, pest control, climate regulation, and availability of water and nutrients, as well as seed/fruit predation, herbivory and competition for pollinators, and considering that these services and dis-services are scale dependent (Zhang *et al.*, 2007).

Summary and Conclusion

Many landscapes worldwide are dominated by agriculture, and large areas of natural forests are disappearing. Human population growth puts further pressures on forests in the search for livelihoods, resources, agricultural land and profits (see Chapter 9, this volume). The land-sparing versus land-sharing debate has evolved in response to these challenges, engaging scientists to contribute research and advocate either land sharing or land sparing as the better way forward.

We have summarized case studies showing that land sharing with agroforestry systems but also multipurpose forests has potential to conserve biodiversity without necessarily compromising crop or wood yield. In these systems, there is no simple trade-off between biodiversity and yield, and moderate shade levels seem to be a good strategy for both biodiversity and yield. Research in this field is still preliminary, and only a few studies have addressed the relationship between yield and biodiversity across multiple years. Besides more research across multiple years and study systems, yield of the primary crop should not be the only indicator of revenues, profitability or farmer livelihood. Several other factors should also be considered. First, market prices and market fluctuations can strongly impact on commodity crops in agroforestry systems (Dahlquist *et al.*, 2007; Gordon *et al.*, 2007). Furthermore, multistrata agroforestry systems provide shade tree products in addition to the main crop produced. These alternative or additional products include medicines, fuel, fibres, fruit and timber, and can provide supplemental income to families that at times account for up to 60% of the income from the agroforestry system (Rice and Greenberg, 2000; Tscharntke *et al.*, 2011). Additionally, farmers of agroforestry commodity crops may receive price premiums for certified products (e.g. organic, shade grown) or may receive payments for other ecosystem services such as carbon sequestration or tree planting (Perfecto *et al.*, 2005; Gordon *et al.*, 2007; Philpott *et al.*, 2007). Premiums can affect the relationship between shade or biodiversity and farmer livelihoods, bypassing the direct effects on yield. Thus, both yield and overall revenue and farmer livelihood must be considered to develop conservation strategies in agroforestry systems (Jaffee, 2007; Méndez *et al.*, 2010).

Land sparing covers a wide range of landscape structures. Large-scale land sparing, where forest remnants are extremely isolated in a continuous agricultural matrix, compromises the exchange and spillover of species between habitats and reduces the delivery of ecosystem services like carbon sequestration, pollination, pest control and water retention to agricultural crops (see also Chapters 4 and 8, this volume). We have shown that, for pollination, functionally important species spillover from forests and other natural habitats to agriculture can positively influence crop yield. Small-scale land sparing where forests and agricultural fields are closely intermixed should therefore be favoured, and additional management strategies that include habitat connectivity for functionally important species should be integrated.

Key socio-economic aspects such as education or women's rights are rarely included in agroecological research, whereas they have been shown to have great impacts on human livelihoods by reducing malnutrition and poverty (Davis, 2002; De Schutter, 2009), as well as boosting agricultural production (FAO, 2006; Ghosh, 2010). Future research should not just evaluate two end points of the land-sparing versus land-sharing concept but should investigate land-use and landscape management practices that combine contrasting concepts in a holistic context-specific manner. If the land-sparing versus land-sharing debate is not transcended, we may find our research supporting alternatives that are not effective at conserving biodiversity or increasing human livelihood (Chappell and LaValle, 2011; Tscharntke *et al.*, 2012).

References

Angelsen, A. and Kaimowitz, D. (1999) Rethinking the causes of deforestation: lessons from economic models. *World Bank Research Observer* 14, 73–98.

Angelsen, A. and Kaimowitz, D. (2001) *Agricultural Technologies and Tropical Deforestation*. CABI Publishing, Wallingford, UK.

Aratrakorn, S., Thunhikorn, S. and Donald, P.F. (2006) Changes in bird communities following conversion of lowland forest to oil palm and rubber plantations in southern Thailand. *Bird Conservation International* 16, 71–82.

Bande, M.M., Grenz, J., Asio, V.B. and Sauerborn, J. (2013) Fiber yield and quality of abaca (*Musa textilis* var. Laylay) grown under different shade conditions, water and nutrient management. *Industrial Crops and Products* 42, 70–77.

Bauhus, J., Pokorny, B., van der Meer, P., Kanowski, P. and Kanninen, M. (2010) Ecosystem goods and services – the key for sustainable plantation. In: Bauhus, J., van der Meer, P. and Kanninen, M. (eds) *Ecosystem Goods and Services from Plantation Forests*. Earthscan, London, pp. 205–227.

Beier, P. and Noss, R.F. (1998) Do habitat corridors provide connectivity? *Conservation Biology* 12, 1241–1252.

Benedick, S., Hill, J.K., Mustaffa, N., Chey, V.K., Maryati, M., Searle, J.B., Schilthuizen, M. and Hamer, K.C. (2006) Impacts of forest fragmentation on butterflies in northern Borneo: species richness, turnover and the value of small fragments. *Journal of Applied Ecology* 43, 967–977.

Bhagwat, S.A., Kushalappa, C.G., Williams, P.H. and Brown, N.D. (2005) The role of informal protected areas in maintaining biodiversity in the Western Ghats of India. *Ecology and Society* 10, 8.

Bloom, D.E. (2011) 7 billion and counting. *Science* 333, 562–569.

Blüthgen, N. and Klein, A.M. (2011) Functional complementarity and specialisation: the role of biodiversity in plant–pollinator interactions. *Basic and Applied Ecology* 12, 282–291.

Boreux, V.E., Krishnan, S., Cheppudira, K.G. and Ghazoul, J. (2012) Impact of forest fragments on bee visits and fruit set in rain-fed and irrigated coffee agro-forests. *Agriculture, Ecosystems and Environment* 172, 42–48.

Brittain, C., Kremen, C. and Klein, A.M. (2012) Biodiversity buffers pollination in changing environmental conditions. *Global Change Biology* 19, 540–547.

Brittain, C., Williams, N., Kremen, C. and Klein, A.M. (2013) Synergistic effects of non-*Apis* bees and honey bees for pollination services. *Proceedings of The Royal Society B: Biological Sciences* 280, 20122767.

Castelletta, M., Sodhi, N.S. and Subaraj, R. (2000) Heavy extinctions of forest-dependent avifauna in Singapore: lessons for biodiversity conservation in Southeast Asia. *Conservation Biology* 14, 1870–1880.

Chappell, M.J. and LaValle, L.A. (2011) Livelihood security and biodiversity: can we have both? An agroecological analysis. *Agriculture and Human Values* 28, 3–26.

Clough, Y., Barkmann, J., Juhrbandt, J., Kessler, M., Wanger, T.C., Anshary, A., Buchori, D., Cicuzza, D., Darras, K., Putra, D.D., Erasmi, S., Pitopang, R. *et al.* (2011) Combining high biodiversity with high yields in tropical agroforests. *Proceedings of the National Academy of Sciences USA* 108, 8311–8316.

Dahlquist, R.M., Whelan, M.P., Winowiecki, L., Polidoro, B., Candela, S., Harvey, C.A., Wulfhorst, J.D., McDaniel, P.A. and Bosque-Perez, N.A. (2007) Incorporating livelihoods in biodiversity conservation: a case study of cacao agroforestry systems in Talamanca, Costa Rica. *Biodiversity and Conservation* 16, 2311–2333.

Davis, M. (2002) *Late Victorian Holocausts: El Niño Famines and the Making of the Third World*. Verso, London/New York.

De Schutter, O. (2009) The right to food and the political economy of hunger. In: *26th McDougall Memorial Lecture. Opening of the 36th Session of the FAO Conference*. Food and Agriculture Organization of the United Nations, Rome.

Dorrough, J., Moll, J. and Crosthwaite, J. (2007) Can intensification of temperate Australian livestock production systems save land for native biodiversity? *Agriculture, Ecosystems and Environment* 121, 222–232.

Egan, J.F. and Mortensen, D.A. (2012) A comparison of land-sharing and land-sparing strategies for plant richness conservation in agricultural landscapes. *Ecological Applications* 22, 459–471.

Ellis, E.C., Goldewijk, K.K., Siebert, S., Lightman, D. and Ramankutty, N. (2010) Anthropogenic transformation of the biomes, 1700 to 2000. *Global Ecology and Biogeography* 19, 589–606.
Eltz, T., Brühl, C.A., Imiyabir, Z. and Linsenmair, K.E. (2003) Nesting and nest trees of stingless bees (Apidae: Milponini) in lowland dipterocarp forests in Sabah, Malaysia, with implications for forest management. *Forest Ecology and Management* 172, 301–313.
Ewers, R.M., Scharlemann, J.P.W., Balmford, A. and Green, R.E. (2009) Do increases in agricultural yield spare land for nature? *Global Change Biology* 15, 1716–1726.
FAO (2006) *The State of Food Insecurity in the World 2006: Eradicating World Hunger – Taking Stock Ten Years After the World Food Summit*. Food and Agriculture Organization of the United Nations, Rome.
FAO (2010) *Global Forest Resources Assessment*. FAO Paper 163. Food and Agriculture Organization of the United Nations, Rome, p. 378.
FAO (2013) *Minimum Dietary Energy Requirement*. FAO Statistics Division, Food and Agriculture Organization of the United Nations, Rome. http://www.fao.org/fileadmin/templates/ess/documents/food_security_statistics/MinimumDietaryEnergyRequirement_en.xls (accessed 29 July 2014).
Ferger, S.W., Bohning-Gaese, K., Wilcke, W., Oelmann, Y. and Schleuning, M. (2013) Distinct carbon sources indicate strong differentiation between tropical forest and farmland bird communities. *Oecologia* 171, 473–486.
Fischer, J., Brosi, B., Daily, G.C., Ehrlich, P.R., Goldman, R., Goldstein, J., Lindenmayer, D.B., Manning, A.D., Mooney, H.A., Pejchar, L., Ranganathan, J. and Tallis, H. (2008) Should agricultural policies encourage land sparing or wildlife-friendly farming? *Frontiers in Ecology and the Environment* 6, 382–387.
Fischer, J., Batary, P., Bawa, K.S., Brussaard, L., Chappell, M.J., Clough, Y., Daily, G.C., Dorrough, J., Hartel, T., Jackson, L.E., Klein, A.M., Kremen, C. *et al.* (2011) Conservation: limits of land sparing. *Science* 334, 593.
Foster, J.B. (2009) *The Ecological Revolution: Making Peace with the Planet*. Monthly Review Press, New York.
Gardiner, M.M., Landis, D.A., Gratton, C., Difonzo, C.D., O'Neal, M., Chacon, J.M., Wayo, M.T., Schmidt, N.P., Mueller, E.E. and Heimpel, G.E. (2009) Landscape diversity enhances biological control of an introduced crop pest in the north-central USA. *Ecological Applications* 19, 143–154.
Garibaldi, L.A., Aizen, M.A., Klein, A.M., Cunningham, S.A. and Harder, L.D. (2011a) Global growth and stability of agricultural yield decrease with pollinator dependence. *Proceedings of the National Academy of Sciences USA* 108, 5909–5914.
Garibaldi, L.A., Steffan-Dewenter, I., Kremen, C., Morales, J.M., Bommarco, R., Cunningham, S.A., Carvalheiro, L.G., Chacoff, N.P., Dudenhoeffer, J.H., Greenleaf, S.S., Holzschuh, A., Isaacs, R. *et al.* (2011b) Stability of pollination services decreases with isolation from natural areas despite honey bee visits. *Ecology Letters* 14, 1062–1072.
Garibaldi, L.A., Steffan-Dewenter, I., Winfree, R., Aizen, M.A., Bommarco, R., Cunningham, S.A., Kremen, C., Carvalheiro, L.G., Harder, L.D., Afik, O., Bartomeus, I., Benjamin, F. *et al.* (2013) Wild pollinators enhance fruit set of crops regardless of honey-bee abundance. *Science* 339, 1608–1611.
Ghosh, J. (2010) *Poverty Reduction in China and India: Policy Implications of Recent Trends*. Working Paper No. 2, United Nations Department of Economic and Social Affairs, New York.
Gilbert, F., Gonzalez, A. and Evans-Freke, I. (1998) Corridors maintain species richness in the fragmented landscapes of a microecosystem. *Proceedings of The Royal Society B: Biological Sciences* 265, 577–582.
Gordon, C., Manson, R., Sundberg, J. and Cruz-Angon, A. (2007) Biodiversity, profitability, and vegetation structure in a Mexican coffee agroecosystem. *Agriculture Ecosystems and Environment* 118, 256–266.
Green, R.E., Cornell, S.J., Scharlemann, J.P.W. and Balmford, A. (2005) Farming and the fate of wild nature. *Science* 307, 550–555.
Guevara, S. and Laborde, J. (1993) Monitoring seed dispersal at isolated standing trees in tropical pastures: consequences for local species availability. *Vegetatio* 108, 319–338.
Gustafsson, L., Baker, S.C., Bauhus, J., Beese, W.J., Brodie, A., Kouki, J., Lindenmayer, D.B., Lõhmus, A., Martínez Pastur, G., Messier C., Neyland, M., Palik, B. *et al.* (2012) Retention forestry to maintain multifunctional forests: a world perspective. *Bioscience* 62, 633–645.
Gutiérrez-Vélez, V.H., Defries, R., Pinedo-Vasquez, M., Uriarte, M., Padoch, C., Baethgen, W., Fernandes, K. and Lim, Y. (2011) High-yield oil palm expansion spares land at the expense of forests in the Peruvian Amazon. *Environmental Research Letters* 6, 044029.
Harvey, C.A., Medina, A., Sanchez, D.M., Vilchez, S., Hernandez, B., Saenz, J.C., Maes, J.M., Casanoves, F. and Sinclair, F.L. (2006) Patterns of animal diversity in different forms of tree cover in agricultural landscapes. *Ecological Applications* 16, 1986–1999.

Hooper, D.U. and Vitousek, P.M. (1998) Effects of plant composition and diversity on nutrient cycling. *Ecological Monographs* 68, 121–149.

Hunter, M.L. and Calhoun, A. (1996) A triad approach to land-use allocation. In: Szaro, R.C. and Johnston, D.W. (eds) *Biodiversity in Managed Landscapes*. Oxford University Press, New York, pp. 477–491.

Jaffee, D. (2007) *Brewing Justice: Fair Trade Coffee, Sustainability, and Survival*. University of California Press, Berkeley, California.

Jasanoff, S. (2004) *States of Knowledge: The Co-production of Science and Social Order*. Routledge, London.

Klein, A.M. (2009) Nearby rainforest promotes coffee pollination by increasing spatio-temporal stability in bee species richness. *Forest Ecology and Management* 258, 1838–1845.

Klein, A.M., Steffan-Dewenter, I. and Tscharntke, T. (2003) Pollination of *Coffea canephora* in relation to local and regional agroforestry management. *Journal of Applied Ecology* 40, 837–845.

Klein, A.M., Steffan-Dewenter, I. and Tscharntke, T. (2006) Rainforest promotes trophic interactions and diversity of trap-nesting hymenoptera in tropical agroforestry. *Journal of Animal Ecology* 75, 315–323.

Klein, A.M., Müller, C., Hoehn, P. and Kremen, C. (2009) Understanding the role of species richness for crop pollination services. In: Naeem, S., Bunker, D.E., Hector, A., Loreau, M. and Perrings, C. (eds) *Biodiversity, Ecosystem Functioning, and Human Wellbeing – an Ecological and Economic Perspective*. Oxford University Press, Oxford, pp. 195–208.

Knoke, T., Stimm, B., Ammer, C. and Moog, M. (2005) Mixed forests reconsidered: a forest economics contribution on an ecological concept. *Forest Ecology and Management* 213, 102–116.

Koh, L.P. and Sodhi, N.S. (2004) Importance of reserves, fragments and parks for butterfly conservation in a tropical urban landscape. *Ecological Applications* 14, 1695–1708.

Koh, L.P. and Wilcove, D.S. (2008) Is oil palm agriculture really destroying tropical biodiversity. *Conservation Letters* 1, 60–64.

Kremen, C., Williams, N.M. and Thorp, R.W. (2002) Crop pollination from native bees at risk from agricultural intensification. *Proceedings of the National Academy of Sciences USA* 99, 16812–16816.

Kremen, C., Williams, N.M., Bugg, R.L., Fay, J.P. and Thorp, R.W. (2004) The area requirements of an ecosystem service: crop pollination by native bee communities in California. *Ecology Letters* 7, 1109–1119.

Larbi, A., Khatib-Salkini, A., Jamal, P.B. and Iniguez, L. (2009) Shrub yield and fodder quality variations in a non-tropical dryland environment in West Asia. *Agroforestry Systems* 75, 147–155.

Layzer, J.A. (2011) *The Environmental Case: Translating Values into Policy*, 3rd edn. CQ Press College, Washington, DC.

Lee, D.R., Ferraro, P.J. and Barrett, C.B. (2001) Introduction: changing perspectives on agricultural intensification, economic development and the environment. In: Lee, D.R. and Barrett, C.B. (eds) *Tradeoffs or Synergies? Agricultural Intensification, Economic Development and the Environment*. CAB International, Wallingford, UK, pp. 1–16.

Levey, D.J., Bolker, B.M., Tewksbury, J.J., Sargent, S. and Haddad, N.M. (2005) Effects of landscape corridors on seed dispersal by birds. *Science* 309, 146–148.

Lindenmayer, D.B. (1994) Wildlife corridor mitigation of logging impacts on fauna in wood-production forests in south-eastern Australia: a review. *Wildlife Research* 21, 323–340.

Lindenmayer, D.B. and Franklin, J.F. (2002) *Conserving Forest Biodiversity. A Comprehensive Multiscaled Approach*. Island Press, Washington DC.

Lindenmayer, D.B., Franklin, J.F., Lõhmus, A., Baker, S.C., Bauhus, J., Beese, W., Brodie, A., Kiehl, B., Kouki, J., Martínez Pastur, G., Messier, C. *et al.* (2012) A major shift to the retention approach for forestry can help resolve some global forest sustainability issues. *Conservation Letters* 5, 421–431.

Lundqvist, J., de Fraiture, C. and Molden, D. (2008) *Saving Water: from Field to Fork – Curbing Losses and Wastage in the Food Chain*. Stockholm International Water Institute Policy Brief, Stockholm.

Mackey, B., Watson, J. and Worboys, G.L. (2008) *Connectivity Conservation and the Great Eastern Ranges Corridor*. Department of Environment and Climate Change, New South Wales, Australia.

Manning, A.D., Fischer, J. and Lindenmayer, D.B. (2006) Scattered trees are keystone structures – implications for conservation. *Biological Conservation* 132, 311–321.

Matson, P.A. and Vitousek, P.M. (2006) Agricultural intensification: will land spared from farming be land spared for nature? *Conservation Biology* 20, 709–710.

Méndez, V.E., Bacon, C.M., Olson, M., Morris, K.S. and Shattuck, A.K. (2010) Agrobiodiversity and shade coffee smallholder livelihoods: a review and synthesis of ten years of research in Central America. *Professional Geographer* 62, 357–376.

Metzger, M.J., Rounsevell, M.D.A., Van den Heiligenberg, H.A.R.M., Pérez-Soba, M. and Soto Hardiman, P. (2010) How personal judgment influences scenario development: an example for future rural development in Europe. *Ecology and Society* 15, 5.

Moguel, P. and Toledo, V.M. (1999) Biodiversity conservation in traditional coffee systems of Mexico. *Conservation Biology* 13, 11–21.

Naeem, S., Bunker, D.E., Hector, A., Loreau, M. and Perrings, C. (2009) *Biodiversity, Ecosystem Functioning, and Human Wellbeing – an Ecological and Economic Perspective*. Oxford University Press, Oxford.

Noss, R.F. (1987) Corridors in real landscapes: a reply to Simberloff and Cox. *Conservation Biology* 1, 159–164.

Noss, R.F. and Beier, P. (2000) Arguing over little things: response to Haddad *et al. Conservation Biology* 14, 1546–1548.

Pardini, R., Marques de Souza, S., Braga-Neto, R. and Metzger, J.P. (2005) The role of forest structure, fragment size and corridors in maintaining small mammal abundance and diversity in an Atlantic forest landscape. *Biological Conservation* 124, 253–266.

Pelletier, N. (2010) Of laws and limits: an ecological economic perspective on redressing the failure of contemporary global environmental governance. *Global Environmental Change* 20, 220–228.

Perfecto, I. and Vandermeer, J. (2010) The agroecological matrix as alternative to the land-sparing/agriculture intensification model. *Proceedings of the National Academy of Sciences USA* 107, 5786–5791.

Perfecto, I., Rice, R.A., Greenberg, R. and Vandervoort, M.E. (1996) Shade coffee: a disappearing refuge for biodiversity. *BioScience* 46, 598–608.

Perfecto, I., Mas, A., Dietsch, T. and Vandermeer, J. (2003) Conservation of biodiversity in coffee agroecosystems: a tri-taxa comparison in southern Mexico. *Biodiversity and Conservation* 12, 1239–1252.

Perfecto, I., Vandermeer, J., Mas, A. and Pinto, L.S. (2005) Biodiversity, yield, and shade coffee certification. *Ecological Economics* 54, 435–446.

Phalan, B., Onial, M., Balmford, A. and Green, R.E. (2011) Reconciling food production and biodiversity conservation: land sharing and land sparing compared. *Science* 333, 1289–1291.

Philpott, S.M., Bichier, P., Rice, R. and Greenberg, R. (2007) Field-testing ecological and economic benefits of coffee certification programs. *Conservation Biology* 21, 975–985.

Philpott, S.M., Arendt, W.J., Armbrecht, I., Bichier, P., Diestch, T.V., Gordon, C., Greenberg, R., Perfecto, I., Reynoso-Santos, R. and Soto-Pinto, L. (2008) Biodiversity loss in Latin American coffee landscapes: review of the evidence on ants, birds, and trees. *Conservation Biology* 22, 1093–1105.

Philpott, S.M., Soong, O., Lowenstein, J.H., Pulido, A.L., Lopez, D.T., Flynn, D.F.B. and Declerck, F. (2009) Functional richness and ecosystem services: bird predation on arthropods in tropical agroecosystems. *Ecological Applications* 19, 1858–1867.

Rice, R.A. and Greenberg, R. (2000) Cacao cultivation and the conservation of biological diversity. *Ambio* 29, 167–173.

Ricketts, T.H. (2004) Tropical forest fragments enhance pollinator activity in nearby coffee crops. *Conservation Biology* 18, 1262–1271.

Roberts, L. (2011) 9 billion? *Science* 333, 540–543.

Romero-Alvarado, Y., Soto-Pinto, L., Garcia-Barrios, L. and Barrera-Gaytan, J.F. (2002) Coffee yields and soil nutrients under the shades of *Inga* sp. vs. multiple species in Chiapas, Mexico. *Agroforestry Systems* 54, 215–224.

Rudel, T.K., Schneider, L., Uriarte, M., Turner, B.L. II, Defries, R., Lawrence, D., Geoghegan, J., Hecht, S.B., Ickowitz, A., Lambin, E.F., Birkenholtz, T., Baptista, S. and Grau, R. (2009) Agricultural intensification and changes in cultivated areas, 1970–2005. *Proceedings of the National Academy of Sciences* 106, 20675–20680.

Santana Sousa, E.H., Batista Matos, M.C., Almeida, R.S. and Teodoro, A.V. (2011) Forest fragments' contribution to the natural biological control of *Spodoptera frugiperda* Smith (Lepidoptera: Noctuidae) in maize. *Brazilian Archives of Biology and Technology* 54, 755–760.

Sekercioglu, C.H., Loarie, S.R., Brenes, F.O., Ehrlich, P.R. and Daily, G.C. (2007) Persistence of forest birds in the Costa Rican agricultural countryside. *Conservation Biology* 21, 482–494.

Seufert, V., Ramankutty, N. and Foley, J.A. (2012) Comparing the yields of organic and conventional agriculture. *Nature* 485, 229–232.

Shrader-Frechette, K. (1995a) Practical ecology and foundations for environmentals ethics. *Journal of Philosophy* 92, 621–635.

Shrader-Frechette, K. (1995b) Evaluating the expertise of experts. *Risk* 6, 115–121.

Simberloff, D.A., Farr, J.A., Cox, J. and Mehlman, D.W. (1992) Movement corridors: conservation bargains or poor investments? *Conservation Biology* 6, 493–504.

Sisi, D.E., Karimi, A.N., Pourahmasi, K. and Taghiyari, H.R. (2012) The effects of agroforestry practices on fiber attributes in *Populus nigra* var. *betulifolia. Trees, Structure and Functions* 26, 435–441.

Smil, V. (2001) *Feeding the World: A Challenge for the 21st Century*. MIT Press, Cambridge, Massachusetts.

Smith, L.C., El Obeid, A.E. and Jensen, H.H. (2000) The geography and causes of food insecurity in developing countries. *Agricultural Economics* 22, 199–215.

Soto-Pinto, L., Perfecto, I., Castillo-Hernandez, J. and Caballero-Nieto, J. (2000) Shade effect on coffee production at the northern Tzeltal zone of the state of Chiapas, Mexico. *Agriculture Ecosystems and Environment* 80, 61–69.

Steffan-Dewenter, I., Kessler, M., Barkmann, J., Bos, M.M., Buchori, D., Erasmi, S., Faust, H., Gerold, G., Glenk, K., Gradstein, S.R., Guhardja, E., Harteveld, M. *et al.* (2007) Tradeoffs between income, biodiversity, and ecosystem functioning during tropical rainforest conversion and agroforestry intensification. *Proceedings of the National Academy of Sciences USA* 104, 4973–4978.

Taki, H., Okabe, K., Yamaura, Y., Matsuura, T., Sueyoshi, M., Makino, S.I. and Maeto, K. (2010) Effects of landscape metrics on *Apis* and non-*Apis* pollinators and seed set in common buckwheat. *Basic and Applied Ecology* 11, 594–602.

Tews, J., Blaum, N. and Jeltsch, F. (2004) Structural and animal species diversity in arid and semi-arid savannas of the southern Kalahari. *Annals of Arid Zone* 42, 1–13.

Tscharntke, T., Clough, Y., Bhagwat, S.A., Buchori, D., Faust, H., Hertel, D., Holscher, D., Juhrbandt, J., Kessler, M., Perfecto, I., Scherber, C., Schroth, G. *et al.* (2011) Multifunctional shade-tree management in tropical agroforestry landscapes – a review. *Journal of Applied Ecology* 48, 619–629.

Tscharntke, T., Clough, Y., Wanger, T.C., Jackson, L., Motzke, I., Perfecto, I., Vandermeer, J.H. and Whitbread, A. (2012) Global livelihood security, biodiversity conservation and the future of agricultural intensification. *Biological Conservation* 151, 53–59.

Uezu, A., Metzger, J.P. and Vielliard, J.M.E. (2005) Effects of structural and functional connectivity and patch size on the abundance of seven Atlantic forest bird species. *Biological Conservation* 123, 507–519.

Van Bael, S., Philpott, S.M., Greenberg, R., Bichier, P., Barber, N.A., Mooney, K.A. and Gruner, D.S. (2008) Birds as predators in tropical agroforestry systems. *Ecology* 89, 928–934.

Waggoner, P.E. (1996) How much land can ten billion people spare for nature? *Daedalus* 125, 73–93.

Waldron, A., Justicia, R., Smith, L. and Sanchez. M. (2012) Conservation through chocolate: a win–win for biodiversity and farmers in Ecuador's lowland tropics. *Conservation Letters* 5, 213–221.

Watson, J.C., Wolf, A.T. and Ascher, J.S. (2011) Forested landscapes promote richness and abundance of native bees (Hymenoptera: Apoidea: Anthophila) in Wisconsin apple orchards. *Environmental Entomology* 40, 621–632.

Whittingham, M.J. (2011) The future of agri-environment schemes: biodiversity gains and ecosystem service delivery? *Journal of Applied Ecology* 48, 509–513.

Wood, G.A.R. and Lass, R.A. (2001) *Cocoa*, 4th edn. Blackwell Science, Oxford.

Wright, H.L., Lake, I.R. and Dolman, P.M. (2012) Agriculture – a key element for conservation in the developing world. *Conservation Letters* 5, 11–19.

Zhang, W., Ricketts, T.H., Kremen, C., Carney, K. and Swinton, S.M. (2007) Ecosystem services and disservices to agriculture. *Ecological Economics* 64, 253–260.

8 Forest Fragmentation and Ecosystem Function

Kelvin S.-H. Peh,[1,2]* Yangchen Lin,[1] Sarah H. Luke,[1] William A. Foster[1] and Edgar C. Turner[1]
[1] *University of Cambridge, Cambridge, UK;* [2] *University of Southampton, Southampton, UK*

Abstract

Urbanization and the spread of agriculture have resulted in high levels of forest loss and fragmentation in many regions of the world. The impact of this anthropogenic fragmentation on ecological functioning has been poorly studied. This chapter assesses the responses of selected plant- and animal-mediated processes to forest fragmentation. Although the number of studies concentrating on different processes is highly variable, with some processes (e.g. nutrient cycling and dung removal) being relatively better studied than others (e.g. predation and parasitism), tentative general insights from recent research include the following: (i) the severity and direction of fragmentation effects are influenced by the individual species involved, the environment and other contributory factors (e.g. surrounding matrix habitat quality); (ii) while processes heavily influenced by larger species are likely to respond negatively to fragmentation, others influenced by a wider range of biotic and abiotic factors are less predictable; (iii) trophic cascades may result in the impacts on one trophic level affecting other parts of the ecosystem; and (iv) the impacts of fragmentation are often exacerbated by synergistic effects such as interactions with the matrix and increased hunting pressures within fragments. Understanding the nature of ecosystem functioning and how functioning becomes disrupted with habitat conversion by combining spatial network models and long-term experimental studies will allow us to predict the eventual trajectories of ecosystems and determine possible remedial actions to preserve healthy functioning ecosystems.

Fragmentation as a Formidable Driver of Change

Habitat loss is among the main drivers of global environmental change, leading to widespread fragmentation of ecosystems (Morris, 2010). Investigations into the ecological effects of fragmentation on forested landscapes, particularly those in the tropics, have been the focus of much conservation research (Bennett and Saunders, 2010). One reason for this attention is that forests usually hold a large number of species. Healthy populations of these species are collectively critical for supporting human well-being, not just by providing harvestable goods like timber and other non-timber forest products but also by delivering a suite of less tangible services such as stabilizing local and global climate, and regulating hydrological flows and cycling nutrients.

*kelvin.peh@gmail.com

There is a general consensus that fragmentation has a strong negative impact on biodiversity within forests (see Chapter 4, this volume). However, the severity and even direction of this impact varies between the species and habitats under investigation (Morris, 2010). The species richness of a fragment is generally related to its size and isolation from the mainland or continuous habitat, in accordance with the theory of island biogeography (Laurance, 2008a). If there is no mainland habitat, the distribution of each individual species is predicted by the distribution of the network of patches, in line with the metapopulation theory (Laurance, 2008a). Populations of species in fragmented habitats may become extinct when the amount of their remaining habitat falls below a critical threshold level (non-linear relationship with habitat loss), which is set by the ecological attributes and extinction parameters of each species (Bennett and Saunders, 2010). Whether species persist in a fragment also depends on direct and indirect biotic interactions (i.e. the metacommunity theory) and interactions with the physical environment (Morris, 2010).

Understanding the patterns of forest species occurrence and the processes underlying them has proven to be challenging because of the idiosyncratic responses of species to habitat fragmentation (see Chapter 4, this volume). It is therefore clear that new approaches must be developed to determine which forest species are likely to be most affected by fragmentation. This is the first step towards understanding the effects of landscape change on forest ecosystem processes, which can then be used to inform and develop cost-effective management strategies to optimize ecosystem function and resilience.

Why Revisit Fragmentation Effects?

In the past decade, there has been a burgeoning interest in forest fragmentation (Fig. 8.1).

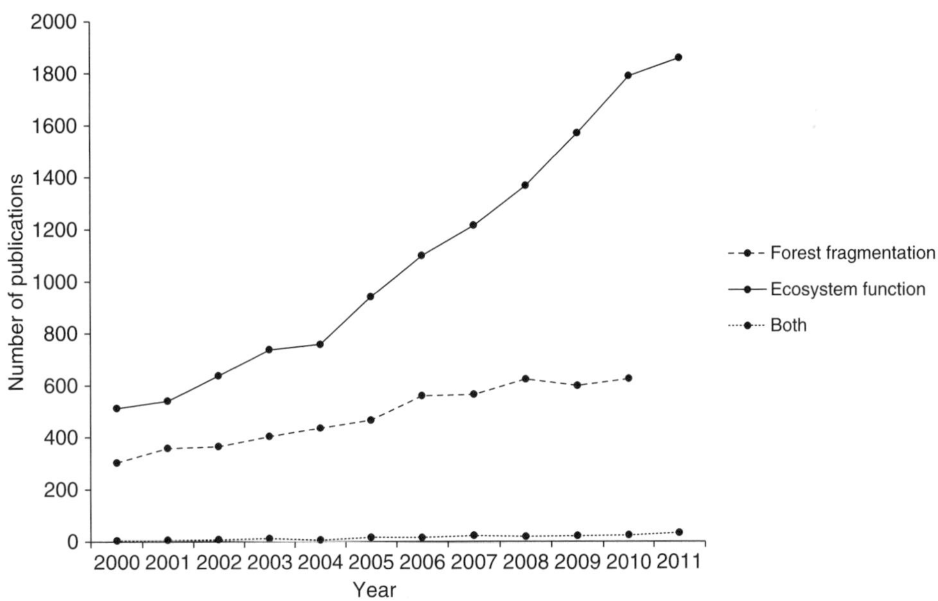

Fig. 8.1. Numbers of research articles found using the search terms 'forest fragmentation' (dashed line), 'ecosystem function' (solid line) and (iii) 'forest fragmentation and ecosystem function' (dotted line) (a total of 5593, 13,025 and 186 papers, respectively) over an 11-year period (2000–2011). These numbers suggest a paucity of ecological information on ecological processes in fragmented forest. Data were obtained from article searches using the Web of Science search engine.

However, until recently, empirical analyses have focused on forest biodiversity and have largely overlooked forest ecosystem processes both within fragments and at the landscape level. Forest fragmentation is commonly believed to affect ecological processes negatively, but empirical evidence for the impact of fragmentation on ecosystem functioning is scarce and equivocal. Moreover, empirical estimations of the degree to which a particular taxon contributes to ecological processes are often highly uncertain (Morris, 2010), making it difficult to determine the precise relationship between diversity and ecosystem functioning (Box 8.1).

Recent advances in ecosystem function research, together with on-going controversies about the direction of fragmentation effects on ecosystem functioning, indicate that it is high time for a thorough review of the subject (Laurance *et al.*, 2001; Morris, 2010). To provide an overview of current forest fragmentation and ecosystem function literature, we performed a literature search using the scientific search engine ISI Web of Science^{SM}. We screened papers published between 2000 and 2011 using the search terms 'forest fragmentation', 'ecosystem function' and 'forest fragmentation AND ecosystem function'. Over the past decade, although the number of publications focusing on ecosystem functioning and forest fragmentation has grown substantially (Fig. 8.1), it appears that only 3% of forest fragmentation studies have included research on ecosystem functioning, demonstrating a continuing paucity of information on the subject.

Subsequent in-depth examination of the literature demonstrated that some processes are better understood than others: nutrient cycling, dung removal, pollination and seed dispersal have been relatively well studied but predation and parasitism much less so. Very few studies have simultaneously investigated the impacts of fragmentation on multiple interacting ecosystem functions, although it is clear that the severity and even direction of impacts can vary widely (Schleuning *et al.*, 2011) and that these variations may be due to the interaction of functions. Multitrophic effects have also received little attention, even though trophic cascades could magnify the impacts of habitat fragmentation (e.g. Allan *et al.*, 2003). It is also important to note that it is not just the average level of functions that may change but also their stability, as fragments may experience hyperdynamic processes (Laurance, 2002).

Here, we review recent evidence on the impacts of fragmentation on ecosystem functions, including biomass productivity, nutrient cycling, predation and parasitism, pollination, seed dispersal, infection rate and aquatic functioning. We concentrate on recent research, as this makes up the bulk of the literature in this growing research field. For example, when we input the terms 'fragmentation' and 'ecosystem function' into Web of Science^{SM}, around 452 of the 557 publications listed (approximately 81%) were published in the last 10 years.

Impact on Ecological Functioning

Biomass productivity

Fragmentation is likely to have significant impacts on forest biomass accumulation, although the effects will depend on the size and placement of the remnant patches (Collinge, 1996). Fragment size is important for biomass productivity because of its effects on abundance and/or turnover of individual species (Laurance, 2008a) and the impacts of edge effects (e.g. Williams-Linera, 1990; Esseen, 1994). The interior of relatively large fragments tends to change less following fragmentation than forest margins, perhaps because edge effects, such as altered microclimate, do not penetrate into the forest interior (Laurance, 2008a). However, further division of these forests into smaller, isolated patches causes the microclimate and vegetation in the fragments to change. In a fragmented Podocarp/broadleaf forest in New Zealand, Young and Mitchell (1994) observed that gross microclimatic edge effects penetrated approximately 50 m from the edge regardless of fragment size. Fragments of less than 9 ha were therefore dominated by edge patterns and processes, and those of less than 1 ha did not contain forest interior microclimatic conditions. It is therefore likely that productivity is stable in

Box 8.1. A brief history of biodiversity–functioning research

The relationship between species diversity and ecosystem functioning has been much studied, not just because it has potentially important consequences for conservation but also because of its importance in determining ecosystem services. Early experimental studies indicated that ecosystem functioning increased with the number of species in a system. It was generally thought that this was because complementarity in resource use led to a more complete utilization of the total resource spectrum (e.g. Tilman, 1999a,b). Many of these early studies assembled communities at random from a species pool. Although such random sampling depicts a possible scenario for gradual loss of diversity (Loreau *et al.*, 2001, 2002), it could potentially introduce a bias, whereby dominant, functionally important species are introduced with increasing diversity (sampling effects), leading to an increase in average primary production (Loreau *et al.*, 2001). If this were the case, ecosystem functioning might increase with diversity, but because of the presence of more productive species rather than because of a better use of resources through the niche complementarity of different species.

Recent evidence suggests that sampling effects, niche complementarity and other underlying mechanisms are not mutually exclusive and act concurrently to determine ecosystem functioning. For example, enhanced functioning in a more diverse synthetic grassland community was shown to be due to both dominant species and a particular combination of complementary species within the community (Tilman *et al.*, 2001). Similarly, diversity-mediated effects on biomass accumulation, nutrient retention and primary productivity have been shown to remain significant after controlling for the strong effects of certain dominant species, indicating that other community-level impacts are important (Hooper *et al.*, 2005). Finally, complementarity has been shown to occur when combinations of species belonged to different functional groups (Loreau and Hector, 2001).

The situation becomes more complex when we consider the impact of different levels of organization of diversity on functioning. While 'species diversity' concerns species richness in terms of the number or relative abundance of individual species, 'functional group diversity' concerns the species composition where groups of species with similar effects on a specific ecosystem-level process (i.e. similar functional traits) are clustered together (Hooper *et al.*, 2005). These different levels of diversity can independently influence ecosystem functioning (Reich *et al.*, 2004). Curiously, both species and functional richness effects have been found not to be dependent on specific species, functional groups or function group combinations in some studies (e.g. Reich *et al.*, 2004), indicating that niche complementarity may be a contributing factor in the positive effect of diversity on functioning.

Most diversity–functioning studies are experimental and control for many confounding effects on the diversity–functioning relationship (Peh and Lewis, 2012). More observational studies in natural ecosystems are needed, as these can incorporate features such as non-random species assemblage and multitrophic influences, which are rarely included in experimental studies (Peh and Lewis, 2012). These 'natural' experiments have been conducted to measure functions such as wood production in Mediterranean forests (Vilà *et al.*, 2007), temperate forests (Caspersen and Pacala, 2001) and tropical tree plantations (Erskine *et al.*, 2006), and have provided evidence of a positive association between species richness and wood production at the regional scale. Few studies have been conducted in tropical forests. However, a meta-analysis by Phillips *et al.* (1994) showed that there was a link between tree species richness and tropical forest dynamics (i.e. tree mortality and recruitment turnover). Bunker *et al.* (2005) simulated 18 possible extinction scenarios using data from a 50 ha tropical forest plot to show that carbon storage in tropical forests depends on species composition. Both these studies demonstrate the importance of diversity in maintaining ecosystem function in tropical forests. Further research that links diversity and function is needed, and studies must go beyond simply describing changes in patterns of diversity with fragmentation.

fragments of more than 9 ha. However, the scale of this edge effect may vary across forest types. For example, Williams-Linera (1990), studying forest edges in Panamanian tropical pre-montane wet forest, found that microclimatic edge effects penetrated only 15 m into the forest, implying that ecological processes, such as tree growth and tree mortality, may be more resilient to edge effects in this forest type.

Generally, any effect of fragmentation on the vegetation composition and structure of a forest may have considerable influence on its net primary productivity (Kelty, 1989). In addition, fragments surrounded by homogeneous, impenetrable matrix tend to be more severely affected by edge effects than those surrounded by heterogeneous, penetrable matrix (such as secondary growth). This is because impacts on the former 'closed' ecosystems are not ameliorated by animals and plants crossing the matrix freely and moving between fragments (Laurance, 2008b; Goldsmith *et al.*, 2011).

Herbivory is thought to act as a strong driving force in plant productivity in fragments, where herbivores tend to benefit from forest edge creation (Tanentzap and Coomes, 2012). The availability of soil nutrients and subsequently net primary productivity seems to increase with the density of herbivores. For example, Feeley and Terborgh (2005) found that above-ground productivity increased with herbivore density on land-bridge islands. However, increased primary productivity may not persist in the longer term, as herbivory may increase the dominance of non-preferred, nutrient-poor plant species and thus decrease nutrient availability over longer time scales (Pastor and Naiman, 1992; Schmitz *et al.*, 2000). Empirical observations made over longer time scales are needed to determine whether there is really a shift in the tree community towards non-preferred species as a result of forest fragmentation.

Could a zero-sum game be possible for fragmentation in terms of ecosystem productivity, such that a decrease in abundance and basal area (indications of decline in biomass) of some species in the forest edge is offset by an equivalent increase in a different suite of species that come to occupy the edge? This has implications for conservation and land management where the aim is often to maintain landscapes capable of high carbon sequestration (Tanentzap and Coomes, 2012). To achieve this, it is necessary to be able to identify a variety of dynamic changes, including shifts in biomass, individual growth rates, abundance, size structure and species composition, and to understand how their relationship changes with the effects of area and isolation over longer time scales. A meta-analysis of 41 data sets of natural ecological communities showed that compensatory dynamics rarely occurred, as species often responded similarly to fluctuations in the overall rates of resource supply (Houlahan *et al.*, 2007). In contrast, a recent analysis of abundance data collected from forest on Barro Colorado Island demonstrated – using a metabolism-based zero-sum framework – compensatory dynamics; thus, a decline in tree abundance was compensated by an increase in tree size in the forest fragment where resources were a limiting factor (Ernest *et al.*, 2009). However, supporting evidence based on longer time series and experimental tests are still lacking.

Nutrient cycling

Nutrients are often limiting resources, and so nutrient cycling of dead organic material – facilitated by the decomposer community – is essential for continued forest productivity. Forest fragmentation may cause changes in a large number of environmental conditions, as well as altering the diversity, abundance and interactions between decomposer organisms. It therefore has the potential to have large impacts on decomposition rates and nutrient cycling within forest ecosystems. However, studies investigating the impacts of fragmentation on decomposition rates show varied effects. For example, a recent study of leaf litter decomposition in a UK woodland found lower rates of decomposition at fragment edges compared with forest interior (Riutta *et al.*, 2012). In contrast, Didham (1998) found much faster decomposition rates at the edge of Amazonian forest fragments. However, a repeat of Didham's experiment by Rubinstein and Vasconcelos (2005), 15 years later, found no differences in decomposition between the forest interior and edge areas. Vasconcelos and Laurance (2005) also found that leaf litter decomposition rates in fragment edges (<100 m from the edge) did not differ from rates in the forest interior (>250 m from the edge, primary forest). No significant difference in rates of decay of aspen stakes was found between different sizes of forest fragment in dry and moist boreal, temperate and tropical forest sites (González *et al.*, 2008).

It seems clear that the combined impact of edge effects, such as increased temperature and light, more wind and decreased moisture, on decomposition is hard to predict and may produce varied results depending on which factor is more important in a particular area. For example, although in some forests abiotic factors such as moisture limitation may reduce decomposition rates in forest edges (Riutta *et al.*, 2012), in other forests, increased tree mortality near forest edges may increase wood availability (Nascimento and Laurance, 2004) and therefore enhance decomposition. Furthermore, trees growing on forest edges may grow at faster rates, producing wood of lower density, which is decomposed faster by saprotrophic fungi (Crockatt, 2012).

A wealth of studies found changes in decomposer assemblages with habitat fragmentation. Although these studies did not measure rates of ecosystem function directly, it is likely that changes in these assemblages would affect decomposition. For example, Rantalainen *et al.* (2005) studied soil decomposer organisms (microbes, fungi and mesofaunal predators) in small humus fragments (between 25 × 25 cm and 236 × 236 cm) in pine forest in Finland and found that species richness of mites and collembolans increased with fragment size.

Several studies have gone further and linked changes in biodiversity with forest fragmentation to changes in function. Genet *et al.* (2001) found lower rates of wood decomposition by termites in fragments than in continuous areas of subtropical dry forest in Puerto Rico. This result did not correlate with overall species richness and abundance; it appears likely that the difference was caused by changes in abundance of three key termite species found mainly in continuous forest. Several studies have considered changes in dung beetle assemblages and rates of dung removal and burial with forest fragmentation. For example, Klein (1989) found significant declines in rates of dung decomposition in Amazon forest fragments compared with intact forest areas. This was correlated with declines in dung beetle species richness, abundance and average beetle size. Other studies also found that dung removal rates increase with functional group richness, but that some groups and species (often large-bodied) play more important roles than others and may be lost more easily from fragments (e.g. Andresen, 2003; Slade *et al.*, 2007; Lewis, 2009). These observations are supported by microcosm experiments, which found that removal of certain dung-processing insect species had large effects on decomposition rates (Rosenlew and Roslin, 2008). Finally, rates of carrion removal may also be influenced by forest fragmentation. Gibbs and Stanton (2001) found that species richness and abundance of carrion beetles was much lower in fragmented forests than in continuous forest in New York State, while in northern Michigan, USA, the rate of carcass burial by burying beetles was significantly greater in larger than in smaller woods (Trumbo and Bloch, 2001). In open and edge habitats, vertebrate scavengers tended to colonize carcasses faster than in the forest interior, where they were more likely to be broken down by microbes and other decomposers (DeVault *et al.*, 2011).

Although there appears to be no universal pattern regarding the effect of fragmentation on decomposition, research certainly indicates that changes in environmental conditions and features of decomposer communities with fragmentation can have significant impacts on nutrient cycling (Gessner *et al.*, 2010). Current research indicates that changes in environmental variables may increase or decrease rates of decomposition, and that such an effect may be system dependent (e.g. Nascimento and Laurance, 2004; Crockatt, 2012; Riutta *et al.*, 2012). Effects of changes in decomposer communities appear idiosyncratic; decomposition rates may be only weakly correlated with total abundance of decomposer organisms (Didham *et al.*, 1996), but some changes may have relatively larger effects on decomposition depending on species identity (e.g. Genet *et al.*, 2001; Lewis, 2009). Additionally, many changes in decomposer communities have been recorded (e.g. Feer and Hingrat, 2005;

Rantalainen *et al.*, 2005), but their impacts on ecosystem function are currently unknown.

Predation

The loss of top predators and consequent shortening of food chains in forest fragments have been widely reported. For example, a recent study in forest fragments on Chiloe, Chile, found lower abundances of all large carnivores studied in smaller forest patches or those with a reduced shape complexity (Farias and Jaksic, 2011). This is related not only to larger home-range requirements and ranging of larger animals but also to increased hunting pressure in fragmented habitats. Such a loss is likely to reduce predation rates, particularly if there is a low redundancy in feeding strategies between species (Farias and Jaksic, 2011).

Changes in the amount of habitat and exposure to matrix habitats may also influence the levels of predation. For example, in a study conducted in deciduous riparian woodland fragments in Idaho, USA, the level of bird nest predation was influenced by the surrounding habitat matrix, with areas adjacent to forest buffers, as opposed to agriculture, experiencing higher predation rates but areas with more agriculture at larger scales also experiencing higher predation. This may be because forest-dwelling predators and wider-habitat predators are both important but operate at different spatial scales (Tewksbury *et al.*, 2006).

Fragmentation also has a profound effect on communities of insectivores and seed predators (Laurance *et al.*, 2001; Ferraz *et al.*, 2003), with species declining more severely in smaller fragments. Insectivorous species of birds (and bats) appear to be more severely affected by fragmentation than some of the other feeding guilds (Laurance *et al.*, 2001; Hill *et al.*, 2011), indicating that insect predation may be reduced in fragments. This has rarely been measured, although a recent study in Kenya found lower levels of antbird predation in forest fragments, probably as a result of the loss of two of the largest species in the antbird flocks (Schleuning *et al.*, 2011). Such losses of predators can also impact on other parts of the ecosystem. For example, Terborgh *et al.* (2001), studying forest islands in Venezuela, found that those without top predators were associated with increased numbers of seed predators and herbivores, higher herbivory, and reduced seedling and sapling recruitment. Higher levels of seedling predation were also found in a study of the fragmented Atlantic Forest of north-eastern Brazil, as a result of the proliferation of leaf-cutter ants (Meyer *et al.*, 2011). In fragmented forests of the Pacific Northwest, an increase in the populations of deer mice (*Peromyscus maniculatus*) led to greater predation of trillium (*Trillium ovatum*) seeds and was associated with lower plant recruitment (Tallmon *et al.*, 2003).

Pollination

Fragmentation has significant impacts on pollination. A review and meta-analysis of the published literature found that plant pollination and reproduction were negatively affected by fragmentation, with animal-pollinated species being more severely impacted than other functional groups (Aguilar *et al.*, 2006). Another recent meta-analysis on bird responses to forest fragmentation revealed that nectivorous birds may be more sensitive to fragmentation than some other feeding guilds (Hill *et al.*, 2011). Brazilian Atlantic Forest fragments contained a lower functional diversity of pollination systems compared with continuous forest, which was also associated with a shift in the functional groups of trees, with functional groups pollinated by birds, flies and non-flying mammals particularly declining. The proportion of hermaphroditic tree species was also higher in fragments in this system (Girão *et al.*, 2007). Similarly, in an Amazonian tree species, higher levels of self-fertilization were also found in fragmented habitats (Dick *et al.*, 2003), indicating that pollination limitation may be important in determining the characteristics of the community.

The characteristics of the matrix seem to be a key factor in determining pollinator dispersal into and from forest fragments.

However, the exact role of the matrix in impacting pollinator dispersal has rarely been studied (Farwig and Berens, 2012). It appears that movement of pollinators may be influenced by the arrangement of forest fragments (Hadley and Betts, 2009) and, in a study that experimentally generated early succession habitats, by the presence of corridors (Tewksbury *et al.*, 2002).

Research on the role of habitat fragments in supporting biodiversity and ecosystem functioning in mosaics of natural and agricultural habitats has proven productive and may be important for crop pollination. For example, higher levels of pollination and productivity have been found in crop habitats with higher bee diversity and functional diversity (Hoehn *et al.*, 2008), which is itself higher closer to forest edges (Klein *et al.*, 2003; Ricketts *et al.*, 2004). In some cases, this may even lead to quantifiable economic benefits of maintaining natural habitat fragments in landscapes to increase productivity and yield quality (Ricketts *et al.*, 2004).

Seed dispersal

Generally, the size and isolation of habitat fragments have been found to reduce fruit removal and seed dispersal both within and between fragments (Farwig and Berens, 2012). Garcia *et al.* (2010), studying birds across a habitat disturbance gradient in three temperate ecosystems, demonstrated the key role that birds play in dispersal and the negative impact of habitat degradation. In Tanzania, the loss of bird species in forest fragments was found to reduce seed dispersal and recruitment of one rainforest tree in forest fragments compared with continuous forest (Cordeiro and Howe, 2003). In northern Spain, a reduction in frugivorous bird diversity as a result of tree loss and fragmentation was associated with a reduction in several components of seed dispersal (García and Martínez, 2012). Seed dispersal in Australian rainforest fragments has been found to decline with the abundance of frugivores (Moran *et al.*, 2009). In the Brazilian Atlantic Forest, smaller forest fragments with lower numbers of larger fauna (probably reduced by hunting pressure) showed lower levels of seed survival, dispersal and recruitment for an endemic palm species (Galetti *et al.*, 2006).

However, such a loss of function is not always the case. For example, although bird abundances differed between fragmented and non-fragmented areas of South African scarp forest, there was no significant change in the estimated fruit removal (Neuschulz *et al.*, 2011). Similarly, in Germany, bird abundance and richness was found to decline with increasing land-use intensity (fragmentation), but visitation rates and seed removal were not significantly affected, possibly due to compensatory increased bird movement (Breitbach *et al.*, 2010).

It is likely that the characteristics of the disperser species affect the impacts of fragmentation on seed dispersal (Spiegel and Nathan, 2007). For example, more mobile bird species fared better than less-mobile species in fragmented forests in East Africa (Farwig and Berens, 2012). The characteristics of the fruiting trees may also influence these impacts. For example, larger-fruited trees are more likely to suffer from short dispersal distances in forest fragments than smaller-fruited trees (McConkey *et al.*, 2012), possibly because of the loss of many of the larger gaped and fruit-feeding species in degraded habitats that are capable of dispersing large seeds (Moran *et al.*, 2004).

As with pollination, matrix quality also has a profound effect on the impacts of fragmentation (Lens *et al.*, 2002). The presence of corridors can also facilitate seed dispersal in fragmented landscapes (Tewksbury *et al.*, 2002). The successful germination of seeds following dispersal may also be impacted by fragmentation. For example, studies investigating the impact of habitat degradation on dung beetles have shown that a loss of dung beetle richness and reduced dung removal is also associated with reduced seed burial (Slade *et al.*, 2011).

Any effect of forest fragmentation on seed dispersers is likely to have important impacts on the distribution and survival of tree species, especially in the tropics, where up to 90% of plant species are dispersed by animals (Howe and Smallwood, 1982).

Any negative impacts of fragmentation on seed dispersers may have even more severe consequences in the future (McConkey *et al.*, 2012) because increased hunting leads to further reductions in large animals (which may well play a disproportionately important role in long distance seed dispersal; Corlett, 2009). In addition, species may need to disperse rapidly to establish themselves in appropriate environmental conditions as the impacts of climate change take effect (Corlett, 2009; Chen *et al.*, 2011).

Infection rate and parasitism

There are still only a limited number of studies investigating the impacts of habitat degradation on host–parasitoid interactions and almost nothing on the impacts of fragmentation, even though these interactions can be crucial for ecosystem stability (Morris, 2010). In the Swiss Plateau, levels of parasitization were lower in fragmented woody patches for trap-nesting wasps, as the parasitoids responded more rapidly to the negative impacts of patch isolation than their hosts (Schüepp *et al.*, 2011). Valladares *et al.* (2012) found host–parasitoid food webs to be reduced in size in smaller fragments in the Chaco Serrano. However, there was also an increase in food-web connectivity, indicating that parasitoids are more generalist in smaller fragments (Valladares *et al.*, 2012).

Loss of top predators in fragmented ecosystems may also influence the incidence of parasitism, with potential consequences for human health. For example, in a study investigating different-sized forest patches in New York State, higher densities of white-footed mice (*Peromyscus leucopus*) in smaller fragments were associated with higher infection rates of ticks carrying Lyme disease, as ticks feeding on mice have a higher likelihood of becoming infected (Allan *et al.*, 2003). The likelihood of humans and other animals becoming infected with Lyme disease is therefore enhanced in smaller fragments.

Levels of brood parasitism in nesting birds may be affected by fragmentation, particularly as a result of interactions with more generalist matrix species (Tewksbury *et al.*, 2006). For example, a weak influence of landscape type on the level of brood parasitism was detected for one species in Illinois, USA (Cottam *et al.*, 2009). In another study of deciduous riparian woodlands in Idaho, USA, the level of brood parasitism was positively influenced by the amount of human habitation in the area and was higher in fragments immediately adjacent to agriculture compared with those next to a forest buffer (Tewksbury *et al.*, 2006). In fragmented landscapes, there is also frequently a sharp increase in plant structural parasites such as lianas, which proliferate at forest edges following fragmentation (Schnitzer and Bongers, 2011) and potentially contribute to heightened tree mortality and turnover (Laurance *et al.*, 1998).

Aquatic functions

Aquatic systems are intrinsically linked to surrounding terrestrial habitats. The riparian zone of rivers and streams, and vegetation surrounding ponds and lakes, regulate inputs of nutrients, organic matter and sediment from the land, help maintain habitat structure, and provide shade and thermal buffering for aquatic communities (e.g. Pusey and Arthington, 2003). Forest fragmentation can cause isolation of pond and pool habitats, cause decreased forest coverage in stream catchments and create gaps in riparian vegetation. Clearance of forest in catchments and beside waterways can increase water temperatures and light inputs, cause loss of in-stream habitat, increase inputs of sediment and change inputs of terrestrial organic matter (e.g. Bunn *et al.*, 1999; Allan, 2004). Despite these large environmental impacts, however, the effect of habitat degradation and fragmentation on aquatic ecosystem function is poorly understood, and studies have so far yielded variable results.

Creation of canopy gaps when riparian forest is lost may stimulate in-stream primary production, which is typically light limited (e.g. Mosisch *et al.*, 2001), while concurrently reducing the supply of terrestrial organic matter to the stream. There may then be a shift in food webs from those dominated by allochthonous (terrestrial) inputs, towards a system dominated by autochthonous (in-stream) production (e.g. England and Rosemond, 2004). This can change the carbon balance of a stream, such that a net consumer (undisturbed stream) (Bunn *et al.*, 1999) becomes a net producer of carbon dioxide. Changes in abiotic conditions, in combination with the abundance of consumer organisms, also affect decomposition of organic matter in streams (Hladyz *et al.*, 2011). Through its effect of increasing temperature, organic pollution, nutrient input and microbial activity, riparian habitat conversion may speed up rates of leaf litter breakdown. However, other consequences of clearance such as sedimentation, reduced pH, heavy metal pollution, channel simplification and loss of macroinvertebrate organisms may slow rates of decomposition (Pascoal *et al.*, 2005; Young *et al.*, 2008), meaning that studies have given mixed results depending on the system being considered. Paul *et al.* (2006) found that rates of leaf decomposition in rivers in the USA were higher at sites with agricultural and urban catchments than at suburban and forested sites, owing to increased microbial and mechanical activity, respectively. However, in a study of streams in Ecuador, rates of decomposition mediated by macroinvertebrates were slower in streams running through pasture than in forested streams, whereas rates mediated by fungal activity showed no change. This effect was caused by a decrease in abundance of a single key decomposer species rather than an overall decline in species richness (Encalada *et al.*, 2010).

Although studies linking fragmentation directly with ecosystem function have been limited, a large number of studies have found changes in diversity and abundance of aquatic organisms with forest fragmentation, which may have significant impacts on functioning, including decomposition and food-web structure. For example, matrix habitat is often a substantial barrier to dispersing amphibians, resulting in a high mortality of dispersing individuals and low recolonization of pools, which may reduce the functional role of these frogs within habitat fragments (Cushman, 2006). Streams in forested catchments in New Zealand had a greater richness and abundance of stream macroinvertebrates than streams in more disturbed catchments (Death and Collier, 2009). Bojsen and Jacobsen (2003) found that macroinvertebrate diversity (alpha and beta) decreased with increasing deforestation, but macroinvertebrate biomass was higher at sites with more open canopies and increased periphyton biomass. Rainforest streams in eastern Madagascar had a higher insect richness than streams with deforested catchments, which were dominated by a few collector–gatherer taxa that could feed on the resulting increased algal growth. Deforested streams lost many of the insects that depend on terrestrial organic matter and were unable to track changes in food availability (Benstead *et al.*, 2003; Benstead and Pringle, 2004). It is common to see shifts in assemblages dominated by Ephemeroptera, Trichoptera and Plecoptera to those dominated by Diptera, molluscs and crustaceans with increasing catchment disturbance (Death and Collier, 2009).

In addition to shifts in assemblage structure, other studies have found changes in behaviour and growth within individual aquatic organisms that may affect their role in aquatic ecosystem functioning. For example, fish are poikilothermic and so their growth and reproduction are affected by changes in water temperature driven by riparian vegetation loss (Pusey and Arthington, 2003). In addition, changes or loss of food supplies from land are likely to affect some taxa. For example, many fish are dependent on terrestrial invertebrates for food, which can make up approximately 85% of the diet of fish in forested streams in Japan. When inputs of insects were experimentally reduced, fish were found to forage elsewhere (Kawaguchi *et al.*, 2003). However, some groups appear able to switch food resources. Ferreira *et al.* (2012) found that characid fish in neotropical streams selected different food sources according to differences in channel structure and stream microhabitat – features that were mediated by riparian forest cover.

Focusing on Multispecies Interactions

Species interactions in forests, especially in the tropics, are poorly understood, and there may be wide-ranging, unforeseen consequences of species loss for food webs and ecosystem functioning (Lewis, 2009). A range of studies has demonstrated the impacts that forest fragmentation can have on species interactions across multiple trophic levels and that there may be both bottom-up and top-down trophic cascades.

Changes in abundance of producers can affect higher trophic levels. For example, fragment edges can experience enhanced plant growth, which then allows population growth of herbivores, especially generalist species (Wirth *et al.*, 2008). Conversely, loss of higher-trophic-level consumers, which may be more sensitive to habitat conversion due to their larger habitat requirements, can have top-down impacts on lower trophic levels. The services provided by these top predators will therefore be lost soon after fragmentation, and may have marked impacts on ecosystem functions such as predation, selective grazing, seed predation and dispersal, disturbance and bioengineering, and nutrient cycling (Duffy, 2003). There will also be consequent impacts further down the food chain, and probably decreases in food-chain length (Duffy, 2003; Dobson *et al.*, 2006). Feer and Hingrat (2005) found that changes in large mammal and particularly howler monkey abundance in fragmented forest in French Guiana had a significant influence on variations in dung beetle species richness. Similar changes in dung beetle community patterns in relation to howler monkey troop occurrence have also been noted in Amazonian (e.g. Andresen, 2003) and Mexican (Estrada *et al.*, 1999) forest fragments. This indirect impact on dung beetle assemblage can also have consequences on ecosystem functions performed by this group. For example, Andresen (2003) also found a higher percentage of seeds buried by beetles in continuous forest than in fragments, which was correlated with the occurrence of dung beetles. Furthermore, 13–19% fewer seeds were buried when large beetles were experimentally excluded, mimicking the effect of forest fragmentation on dung beetles and dung beetle-mediated seed removal. This has potentially important impacts on plant communities, as seeds buried by dung beetles are much less likely to be predated by rats than non-buried seeds (e.g. Andresen, 2003). Fragmentation-induced changes in howler monkeys and dung beetles therefore have potential negative impacts on plant reproductive success.

Changes in environmental conditions and in movement patterns of individuals in fragmented forest can alter interactions and the strength of competition. For example, fragmentation of forest habitat in the USA allowed mesopredators (mainly raccoons and opossums) to discover and remove mouse carcasses more quickly than in contiguous habitat, and so outcompete insects and microbes (DeVault *et al.*, 2011). Changes in patterns of dispersal can also affect the intensity of species interactions (Fagan *et al.*, 1999). Many species do not disperse across gaps and therefore forage parallel to edges. Others actively avoid edges to reduce predation risk, resulting in an unbalanced activity pattern (e.g. Fagan *et al.*, 1999; Restrepo and Vargas, 1999). For example, germination of nuts has been found to be lower at the edges of hickory forest fragments, possibly as a result of reduced squirrel foraging (Sork, 1983). Similarly, seed predators in Virginian woodlands avoid edges, which have an increased abundance of their predators (Bowers and Dooley, 1993), which probably reduces seed predation in these areas.

Synergistic Impacts on Functioning

It is increasingly clear that multiple threats, acting together in complex ways, are responsible for losses of biodiversity and reduced ecosystem functioning globally (Turner, 2005; Brook *et al.*, 2008). The impacts of forest fragmentation on biological communities may be driven by multiple related factors such as a reduction in habitat area, edge effects (which may also act synergistically to drive species loss; Ewers *et al.*, 2007), the degree of isolation and habitat connectivity (Didham *et al.*, 1996). Land-use in adjacent habitat areas, previous land-use (Turner, 2005) and biotic interactions and secondary effects (such as trophic cascades, invasive species and the functional links between species in complex ecosystems) are also highly important

(Lewis, 2009). There are many co-dependent drivers of changes (Gardner *et al.*, 2009).

The exact effect of fragmentation is extremely variable among species and depends on factors such as whether a species is a habitat interior specialist, habitat generalist or edge species, as well as its mobility (Larsen *et al.*, 2005). For example, in Swedish deciduous forests, habitat specialists and clonally reproducing ground-layer plant species are more severely impacted by fragment isolation than are other plant species (Dupré and Ehrlén, 2002). In Venezuela, large-bodied, forest specialist and rare dung beetles were more likely than other groups to become extinct following fragmentation (Larsen *et al.*, 2008).

Landscape-scale habitat characteristics and the wider matrix can also be important in predicting impacts on the biotic community (Mazerolle and Villard, 1999). Ricketts (2001) showed that the degree of matrix permeability for different butterfly species can influence the effective isolation of a fragment. Modelling work also indicates that the configuration of fragments may be important, particularly for poor dispersers and for rarer habitat types (King and With, 2002). Other characteristics and detrimental impacts from the matrix can also directly influence the survival of species in fragments. For example, forest edges can trap airborne nutrients and pollutants originating in adjacent agricultural and urban areas. Edges therefore experience much higher rates of deposition than forest interiors (Weathers *et al.*, 2001), potentially impacting detrimentally on biological communities. In south-east Australian forest fragments, the intrusion of grazing animals into forest fragments from surrounding agricultural areas significantly impacts on several components of diversity and reduces litter decomposition rates in forest edges (Lindsay and Cunningham, 2009). Some of these negative matrix effects can be reduced by forest regrowth, effectively reconnecting fragments for some species and allowing increased use of edges and recolonization (Laurance *et al.*, 2001). Finally, the importance of the matrix may vary among species, depending on dispersal and habitat preference. There is evidence of such variation between feeding guilds of birds, as has been found from a study of bird migration in Peru (Lloyd and Marsden, 2011).

The nature of ecosystem function change following fragmentation is also influenced by the relationship between species responses to habitat change (species disassembly rules and response traits) and the functional roles that species perform in ecosystems (input effect traits; Ostfield and LoGiudice, 2003; Zavaleta *et al.*, 2009), making results difficult to predict and potentially exacerbating impairment of ecosystem functioning. There is evidence from studies on bees and dung beetles that larger species may be more sensitive to habitat fragmentation and also more effective in supporting ecosystem functioning (Larsen *et al.*, 2005). Early loss of these species will therefore have a non-random and disproportionately severe impact on ecosystem functioning. Complex relationships between groups of species could also lead to further synergistic losses of species and impairment of ecosystem functioning.

Promising Avenues for Future Research

Forest fragmentation is inevitable, in both agricultural and urban landscapes. It is therefore important to understand how to manage fragmented *landscapes*, as opposed to just the fragments themselves, if we are to optimize ecosystem function and resilience. This is increasingly important in the face of mounting global change, which may increase the risk of a sudden ecosystem-wide collapse.

Studies investigating the impacts of fragmentation on measured ecosystem processes are still rare. More research needs to concentrate on this functional approach rather than cataloguing changes in biodiversity alone (Didham *et al.*, 1996). Related to this, more work should be carried out on how species and community characteristics (such as abundances and degree of redundancy) and species disassembly rules impact on function relationships (Larsen *et al.*, 2005; Lewis, 2009). We recommend directly monitoring 'indicator' ecosystem processes in both the fragments and the surrounding matrix over time, for both fragmentation and defragmentation processes. It is important to understand how function changes in both directions of landscape change in order to determine optimal management plans and policies.

Because of the complex nature of global change and the large number of correlated synergistic impacts that can affect the biological environment (Brook *et al.*, 2008; Sodhi *et al.*, 2010), the development of experimental approaches that explicitly test the impacts of habitat fragmentation on biodiversity and ecosystem function are essential (Rosenlew and Roslin, 2008; Morris, 2010). These should be of a sufficient scale to embrace a range of different taxa and ecosystem functions (Schleuning *et al.*, 2011) and to allow a quantitative assessment of ecosystem function stability. The value of such experimental projects is clear from their longevity (Laurance *et al.*, 2001) and influence (Gardner *et al.*, 2009) and from the developing interest in new multidisciplinary projects of this kind. For example, the recently launched large-scale tropical forest fragmentation experiment, the Stability of Altered Forest Ecosystems (SAFE) Project (Ewers *et al.*, 2011), allows landscape- and fragment-level effects to be differentiated.

We propose that both the experimental and theoretical challenges can be tackled using the robust and flexible spatial network approach (e.g. Fortuna *et al.*, 2010; Carrara *et al.*, 2012) where nodes represent values of an ecological attribute in different fragments and links represent how the values influence one another. This approach can also be used to study keystone communities (Mouquet *et al.*, 2013), which can enhance ecosystem-level understanding and management in fragmented landscapes. Furthermore, social network models can be used to accurately predict connectivity in fragmented landscapes where data are scarce (Fletcher *et al.*, 2011).

In addition to spatial considerations, biotic interactions can have unexpected non-linear effects and are important for understanding and predicting the effects of global change on ecosystems (Tylianakis *et al.*, 2008). Morris (2010), Fenoglio *et al.* (2012) and Valladares *et al.* (2012) provided empirical examples of how fragmentation profoundly affects the structure of species interaction networks. Some pioneering studies have also attempted to unify spatial and interaction networks, such as those of Sole and Montoya (2006) and Pillai *et al.* (2011).

In addition, the rates of ecological processes like fragmentation, dispersal and fluctuations in species abundances are as important as, if not more important than, snapshot measurements of these quantities. Increasing theoretical attention is being paid to community dynamics (e.g. Kéfi *et al.*, 2012), and time series data are becoming increasingly available (albeit mostly on relatively simple aquatic systems, e.g. Carpenter *et al.*, 2011) to validate the theory. We envision that long-term empirical data from large projects like SAFE (Ewers *et al.*, 2011) can eventually be combined with theory for a better understanding and management of complex ecosystem functioning in fragmented landscapes.

The next big step for fragmentation research is to apply these multidisciplinary approaches to monitor actual developments on the ground so that we can protect functionally complementary species (Elmqvist *et al.*, 2003) or suits of species (Walker, 1995) and ecosystem functioning (Garcia *et al.*, 2010) for the long term. This should include the development of more holistic conservation programmes that are focused on the effective conservation of ecosystem functioning, and not only on the preservation of biodiversity and charismatic species (Garcia *et al.*, 2010; Sodhi *et al.*, 2010; Farwig and Berens, 2012; McConkey *et al.*, 2012).

Conclusion

The impacts of forest fragmentation on biodiversity have been well investigated and can be severe. In general, smaller more isolated fragments lose more species, and larger, less dispersive species are lost more quickly from fragments. The effects of forest fragmentation on ecosystem functioning have been much less investigated and the number of studies concentrating on different processes is highly variable. For example, nutrient cycling, dung removal, pollination and seed dispersal have been relatively well studied but predation and parasitism much less so. Fragmentation frequently has a marked effect on a wide range of ecosystem processes, although the severity and direction of the impacts are influenced by the individual species involved, the environment and contributory factors such as the surrounding matrix habitat quality.

Ecosystem processes influenced by a wider range of biotic and abiotic factors are less predictable in their response to fragmentation compared with those determined by fewer factors. Functions heavily influenced by larger species are particularly likely to be impacted negatively. Impacts on one trophic level may have consequences for other parts of the ecosystem as a result of cascading effects through the trophic levels. The impacts of fragmentation on ecosystem functioning are often exacerbated by synergistic effects such as interactions with the matrix and increased hunting pressure in fragmented forests.

Future work should focus on combining systems-oriented modelling approaches and robust, long-term experimental studies to inform the maintenance of ecosystems for more resilient functioning in the face of inevitable habitat change.

References

Aguilar, R., Ashworth, L., Galetto, L. and Aizen, M.A. (2006) Plant reproductive susceptibility to habitat fragmentation: review and synthesis through a meta-analysis. *Ecology Letters* 9, 968-–980.

Allan, B.F., Keesing, F. and Ostfeld, R.S. (2003) Effect of forest fragmentation on Lyme disease risk. *Conservation Biology* 17, 267–272.

Allan, J. (2004) Landscapes and riverscapes: the influence of land use on stream ecosystems. *Annual Review of Ecology, Evolution, and Systematics* 35, 257–284.

Andresen, E. (2003) Effect of forest fragmentation on dung beetle communities and functional consequences for plant regeneration. *Ecography* 26, 87–97.

Bennett, A.F. and Saunders, D.A. (2010) Habitat fragmentation and landscape change. In: Sodhi, N.S. and Ehrlich, P.R. (eds) *Conservation Biology for All*. Oxford University Press, Oxford, pp. 88–106.

Benstead, J.P. and Pringle, C.M. (2004) Deforestation alters the resource base and biomass of endemic stream insects in eastern Madagascar. *Freshwater Biology* 49, 490–501.

Benstead, J.P., Douglas, M.M. and Pringle, C.M. (2003) Relationships of stream invertebrate communities to deforestation in eastern Madagascar. *Ecological Applications* 13, 1473–1490.

Bojsen, B.H. and Jacobsen, D. (2003) Effects of deforestation on macroinvertebrate diversity and assemblage structure in Ecuadorian Amazon streams. *Archiv für Hydrobiologie* 158, 317–342.

Bowers, M.A. and Dooley, J.L. Jr (1993) Predation hazard and seed removal by small mammals: microhabitat versus patch scale effects. *Oecologia* 94, 247–254.

Breitbach, N., Laube, I., Steffan-Dewenter, I. and Böhning-Gaese, K. (2010) Bird diversity and seed dispersal along a human land-use gradient: high seed removal in structurally simple farmland. *Oecologia* 162, 965–976.

Brook, B.W., Sodhi, N.S. and Bradshaw, C.J.A. (2008) Synergies among extinction drivers under global change. *Trends in Ecology and Evolution* 23, 453–460.

Bunker, D.E., De Clerck, F., Bradford, J.C., Colwell, R.K., Perfecto, I., Phillips, O.L., Sankaran, M. and Naeem, S. (2005) Species loss and aboveground carbon storage in a tropical forest. *Science* 310, 1029–1031.

Bunn, S.E., Davies, P.M. and Mosisch, T.D. (1999) Ecosystem measures of river health and their response to riparian and catchment degradation. *Freshwater Biology* 41, 333–345.

Carpenter, S.R., Cole, J.J., Pace, M.L., Batt, R., Brock, W.A., Cline, T., Coloso, J., Hodgson, J.R., Kitchell, J.F., Seekell, D.A., Smith, L. and Weidel, B. (2011) Early warnings of regime shifts: a whole-ecosystem experiment. *Science* 332, 1079–1082.

Carrara, F., Altermatt, F., Rodriguez-Iturbe, I. and Rinaldo, A. (2012) Dendritic connectivity controls biodiversity patterns in experimental metacommunities. *Proceedings of the National Academy of Sciences USA* 109, 5761–5766.

Caspersen, J.P. and Pacala, S.W. (2001) Successional diversity and forest ecosystem function. *Ecological Research* 16, 895–903.

Chen, I.C., Hill, J.K., Ohlemüller, R., Roy, D.B. and Thomas, C.D. (2011). Rapid range shifts of species associated with high levels of climate warming. *Science* 333, 1024–1026.

Collinge, S.K. (1996) Ecological consequences of habitat fragmentation: implications for landscape architecture and planning. *Landscape and Urban Planning* 36, 59–77.

Cordeiro, N.J. and Howe, H.F. (2003) Forest fragmentation severs mutualism between seed dispersers and an endemic African tree. *Proceedings of the National Academy of Sciences USA* 100, 14052–14056.

Corlett, R.T. (2009) Seed dispersal distances and plant migration potential in tropical east Asia. *Biotropica* 41, 592–598.

Cottam, M.R., Robinson S.K., Heske, E.J., Brawn, J.D. and Rowe, K.C. (2009) Use of landscape metrics to predict avian nest survival in a fragmented midwestern forest landscape. *Biological Conservation* 142, 2464–2475.

Crockatt, M.E. (2012) Are there edge effects on forest fungi and if so do they matter? *Fungal Biology Reviews* 26, 94–101.

Cushman, S.A. (2006) Effects of habitat loss and fragmentation on amphibians: a review and prospectus. *Biological Conservation* 128, 231–240.

Death, R.G. and Collier, K.J. (2009) Measuring stream macroinvertebrate responses to gradients of vegetation cover: when is enough enough? *Freshwater Biology* 55, 1447–1464.

DeVault, T.L., Olson, Z.H., Beasley, J.C. and Rhodes, O.E. Jr (2011) Mesopredators dominate competition for carrion in an agricultural landscape. *Basic and Applied Ecology* 12, 268–274.

Dick, C.W., Etchelecu, G. and Austerlitz, F. (2003) Pollen dispersal of tropical trees (*Dinizia excelsa:* Fabaceae) by native insects and African honeybees in pristine and fragmented Amazonian rainforest. *Molecular Ecology* 12, 753–764.

Didham, R.K. (1998) Altered leaf-litter decomposition rates in tropical forest fragments. *Oecologia* 116, 397–406.

Didham, R.K., Ghazoul, J., Stork, N.E. and Davis, A.J. (1996) Insects in fragmented forests: a functional approach. *Trends in Ecology and Evolution* 11, 255–260.

Dobson, A., Lodge, D., Alder, J., Cumming, G.S., Keymer, J., McGlade, J., Mooney, H., Rusak, J.A., Sala, O., Wolters, V., Wall, D., Winfree, R. and Xenopoulos, M.A. (2006) Habitat loss, trophic collapse, and the decline of ecosystem services. *Ecology* 87, 1915–1924.

Duffy, J.E. (2003) Biodiversity loss, trophic skew and ecosystem functioning. *Ecology Letters* 6, 680–687.

Dupré, C. and Ehrlén, J. (2002) Habitat configuration, species traits and plant distributions. *Journal of Ecology* 90, 796–805.

Elmqvist, T., Folke, C., Nyström, M., Peterson, G., Bengtsson, J., Walker, B. and Norberg, J. (2003) Response diversity, ecosystem change, and resilience. *Frontiers in Ecology and the Environment* 1, 488–494.

Encalada, A.C., Calles, J., Ferreira, V., Canhoto, C.M. and Graçca, M.A.S. (2010) Riparian land use and the relationship between the benthos and litter decomposition in tropical montane streams. *Freshwater Biology* 55, 1719–1733.

England, L.E. and Rosemond, A.D. (2004) Small reductions in forest cover weaken terrestrial–aquatic linkages in headwater streams. *Freshwater Biology* 49, 721–734.

Ernest, S.K.M., White, E.P. and Brown, J.H. (2009) Changes in a tropical forest support metabolic zero-sum dynamics. *Ecology Letters* 12, 507–515.

Erskine, P.D., Lamb, D. and Bristow, M. (2006) Tree species diversity and ecosystem function: can tropical multi-species plantations generate greater productivity? *Forest Ecology and Management* 233, 205–210.

Esseen, P.-A. (1994) Tree mortality patterns after experimental fragmentation of an old-growth conifer forest. *Biological Conservation* 68, 19–28.

Estrada, A., Anzures, D.A. and Coates-Estrada, R. (1999) Tropical rain forest fragmentation, howler monkeys (*Alouatta palliata*), and dung beetles at Los Tuxtlas, Mexico. *American Journal of Primatology* 48, 253–262.

Ewers, R.M., Thorpe, S. and Didham, R.K. (2007) Synergistic interactions between edge and area effects in a heavily fragmented landscape. *Ecology* 88, 96–106.

Ewers, R.M., Didham, R.K., Fahrig, L., Ferraz, G., Hector, A., Holt, R.D., Kapos, V., Reynolds, G., Sinun, W., Snaddon, J.L. and Turner, E.C. (2011) A large-scale forest fragmentation experiment: the Stability of Altered Forest Ecosystems Project. *Philosophical Transactions of The Royal Society B: Biological Sciences* 366, 3292–3302.

Fagan, W.F., Cantrell, R.S. and Cosner, C. (1999) How habitat edges change species interactions. *American Naturalist* 153, 165–182.

Farias, A.A. and Jaksic, F.M (2011) Low functional richness and redundancy of a predator assemblage in native forest fragments of Chiloe Island, Chile. *Journal of Animal Ecology* 80, 809–817.

Farwig, N. and Berens, D.G. (2012) Imagine a world without seed dispersers: a review of threats, consequences and future directions. *Basic and Applied Ecology* 13, 109–115.

Feeley, K.J. and Terborgh, J.W. (2005) The effects of herbivore density on soil nutrients and tree growth in tropical forest fragments. *Ecology* 86, 116–124.

Feer, F. and Hingrat, Y. (2005) Effects of forest fragmentation on a dung beetle community in French Guiana. *Conservation Biology* 19, 1103–1112.

Fenoglio, M.S., Srivastava, D., Valladares, G., Cagnolo, L. and Salvo, A. (2012) Forest fragmentation reduces parasitism via species loss at multiple trophic levels. *Ecology* 93, 2407–2420.

Ferraz, G., Russell, G.J., Stouffer, P.C., Bierregaard, R.O., Pimm, S.L. and Lovejoy, T.E. (2003) Rates of species loss from Amazonian forest fragments. *Proceedings of the National Academy of Sciences USA* 100, 14069–14073.

Ferreira, A., Gerhard, P. and Cyrino, J.E.P. (2012) Diet of *Astyanax paranae* (Characidae) in streams with different riparian land covers in the Passa-Cinco River basin, southeastern Brazil. *Iheringia, Série Zoologia* 102, 80–87.

Fletcher, R.J., Acevedo, M.A., Reichert, B.E., Pias, K.E. and Kitchens, W.M. (2011) Social network models predict movement and connectivity in ecological landscapes. *Proceedings of the National Academy of Sciences USA* 108, 19282–19287.

Fortuna, M.A., Albaladejo, R.G., Fernandez, L., Aparicio, A. and Bascompte, J. (2010) Networks of spatial genetic variation across species. *Proceedings of the National Academy of Sciences USA* 106, 19044–19049.

Galetti, M., Donatti, C.I., Pires, A.S., Guimarães, J.R. and Jordano, P. (2006) Seed survival and dispersal of an endemic Atlantic Forest palm: the combined effects of defaunation and forest fragmentation. *Botanical Journal of the Linnean Society* 151, 141–149.

García, D. and Martínez, D. (2012) Species richness matters for the quality of ecosystem services: a test using seed dispersal by frugivorous birds. *Proceedings of The Royal Society B: Biological Sciences* 279, 3106–3113.

Garcia, D., Zamora, R. and Amico, G.C. (2010) Birds as suppliers of seed dispersal in temperate ecosystems: conservation guidelines from real-world landscapes. *Conservation Biology* 24, 1070–1079.

Gardner, T.A., Barlow, J., Chazdon, R., Ewers, R.M., Harvey, C.A., Peres, C.A. and Sodhi, N.S. (2009) Prospects for tropical forest biodiversity in a human-modified world. *Ecology Letters*, 561–582.

Genet, J.A., Genet, K.S., Burton, T.M., Murphy, P.G. and Lugo, A.E. (2001) Response of termite community and wood decomposition rates to habitat fragmentation in a subtropical dry forest. *Tropical Ecology* 42, 35–49.

Gessner, M.O., Swan, C.M., Dang, C.K., McKie, B.G., Bardgett, R.D., Wall, D.H. and Hättenschwiler, S. (2010) Diversity meets decomposition. *Trends in Ecology and Evolution* 25, 372–380.

Gibbs, J.P. and Stanton, E.J. (2001) Habitat fragmentation and arthropod community change: carrion beetles, phoretic mites, and flies. *Ecological Applications* 11, 79–85.

Girão, L.C., Lopes, A.V., Tabarelli, M. and Bruna, E.M. (2007) Changes in tree reproductive traits reduce functional diversity in a fragmented Atlantic forest landscape. *PLoS One* 2, e908.

Goldsmith, G.R., Comita, L.S. and Chua, S.C. (2011) Evidence for arrested succession within a tropical forest fragment in Singapore. *Journal of Tropical Ecology* 27, 323–326.

González, G., Gould, W.A., Hudak, A.T. and Hollingsworth, T.N. (2008) Decay of aspen (*Populus tremuloides* Michx.) wood in moist and dry boreal, temperate, and tropical forest fragments. *Ambio* 37, 588–597.

Hadley, A.S. and Betts, M.G.(2009) Tropical deforestation alters hummingbird movement patterns. *Biology Letters* 5, 207–210.

Hill, J.K., Gray, M.A., Chey, V.K., Benedick, S., Tawatao, N. and Hamer, K.C. (2011) Ecological impacts of tropical forest fragmentation: how consistent are patterns in species richness and nestedness? *Philosophical Transactions of The Royal Society B: Biological Sciences* 366, 3265–3276.

Hladyz, S., Åbjörnsson, K., Chauvet, E., Dobson, M., Elosegi, A., Ferreira, V., Fleituch, T., Gessner, M.O., Giller, P.S., Gulis, V., Hutton, S.A., Lacoursière, J.O. *et al.* (2011) Stream ecosystem functioning in an agricultural landscape: the importance of terrestrial–aquatic linkages. *Advances in Ecological Research* 44, 211–276.

Hoehn, P., Tscharntke, T., Tylianakis, J.M. and Steffan-Dewenter, I. (2008) Functional group diversity of bee pollinators increases crop yield. *Proceedings of The Royal Society B: Biological Sciences* 275, 2283–2291.

Hooper, D.U., Chaplin, S. III, Ewel, J.J., Hector, A., Inchausti, P., Lavorel, S., Lawton, J.H., Lodge, D.M., Loreau, M., Naeem, S., Schmid, B., Setälä, H. *et al.* (2005) Effects of biodiversity on ecosystem functioning: a consensus of current knowledge. *Ecological Monographs* 75, 3–35.

Houlahan, J.E., Currie, D.J., Cottenie, K., Cumming, G.S., Ernest, S.K., Findlay, C.S., Fuhlendorf, S.D., Gaedke, U., Legendre, P., Magnuson, J.J., McArdle, B.H., Muldavin, E.H. *et al.* (2007) Compensatory dynamics are rare in natural ecological communities. *Proceedings of the National Academy of Sciences USA* 104, 3273–3277.

Howe, H.F. and Smallwood, J. (1982) Ecology of seed dispersal. *Annual Review of Ecology, Evolution, and Systematics* 13, 201–228.

Kawaguchi, Y., Taniguchi, Y. and Nakano, S. (2003) Terrestrial invertebrate inputs determine the local abundance of stream fishes in a forested stream. *Ecology* 84, 701–708.

Kéfi, S., Berlow, E.L., Wieters, E.A., Navarrete, S.A., Petchey, O.L., Wood, S.A., Boit, A., Joppa, L.N., Lafferty, K.D., Williams, R.J., Martinez, N.D., Menge, B.A. *et al.* (2012) More than a meal... integrating non-feeding interactions into food webs. *Ecology Letters* 15, 291–300.

Kelty, M.J. (1989) Productivity of New England hemlock/hardwood stands as affected by species composition and canopy structure. *Forest Ecology and Management* 28, 237–257.

King, A.W. and With, K.A. (2002) Dispersal success on spatially structured landscapes: when do spatial pattern and dispersal behavior really matter? *Ecological Modelling* 147, 23–39.

Klein, A.-M., Steffan-Dewenter, I. and Tscharntke, T. (2003) Fruit set of highland coffee increases with the diversity of pollinating bees. *Proceedings of The Royal Society B: Biological Sciences* 270, 955–961.

Klein, B.C. (1989) Effects of forest fragmentation on dung and carrion beetle communities in central Amazonia. *Ecology* 70, 1715–1725.

Larsen, T.H., Williams, N.M. and Kremen, C. (2005) Extinction order and altered community structure rapidly disrupt ecosystem functioning. *Ecology Letters* 8, 538–547.

Larsen, T.H., Lopera, A. and Forsyth, A. (2008) Understanding trait-dependent community disassembly: dung beetles, density functions, and forest fragmentation. *Conservation Biology* 22, 1288–1298.

Laurance, W.F. (2002) Hyperdynamism in fragmented habitats. *Journal of Vegetation Science* 13, 595–602.

Laurance, W.F. (2008a) Theory meets reality: how habitat fragmentation research has transcended island biogeographic theory. *Biological Conservation* 141, 1731–1744.

Laurance, W.F. (2008b) Theory meets reality in fragmented forests. *Animal Conservation* 11, 364–365.

Laurance, W.F., Ferreira, L.V., Rankin-de Merona, J.M. and Laurance, S.G. (1998) Rain forest fragmentation and the dynamics of Amazonian tree communities. *Ecology* 79, 2032–2040.

Laurance, W.F., Lovejoy, T.E., Vasconcelos, H.L., Bruna, E.M., Didham, R.K., Stouffer, P.C., Gascon, C., Bierregaard, R.O., Laurance, S.G. and Sampaio, E. (2001) Ecosystem decay of Amazonian forest fragments: a 22-year investigation. *Conservation Biology* 16, 605–618.

Lens, L., Van Dongen, S., Norris, K., Githiru, M. and Matthysen, E. (2002) Avian persistence in fragmented rainforest. *Science* 298, 1236–1238.

Lewis, O.T. (2009) Biodiversity change and ecosystem function in tropical forests. *Basic and Applied Ecology* 10, 97–102.

Lindsay, E.A. and Cunningham, S.A. (2009) Livestock grazing exclusion and microhabitat variation affect invertebrates and litter decomposition rates in woodland remnants. *Forest Ecology and Management* 258, 178–187.

Lloyd, H. and Marsden, S.J. (2011) Between-patch bird movements within a High-Andean *Polylepis* woodland/matrix landscape: implications for habitat restoration. *Restoration Ecology* 19, 74–82.

Loreau, M. and Hector, A. (2001) Partitioning selection and complementarity in biodiversity experiments. *Nature* 412, 72–76.

Loreau, M., Naeem, S., Inchausti, P., Bengtsson, J., Grime, J.P., Hector, A., Hooper, D.U., Huston, M.A., Raffaelli, D., Schimid, B., Tilman, D. and Wardle, D.A. (2001) Biodiversity and ecosystem functioning: current knowledge and future challenges. *Science* 294, 804–808.

Loreau, M., Naeem, S. and Inchausti, P. (2002) *Biodiversity and Ecosystem Functioning: Synthesis and Perspectives*. Oxford University Press, Oxford.

Mazerolle, M.J. and Villard, M.A. (1999) Patch characteristics and landscape context as predictors of species presence and abundance: a review. *Ecoscience* 6, 117–124.

McConkey, K.R., Prasad, S., Corlett, R.T., Campos-Arceiz, A., Brodie, J.F., Rogers, H. and Santamaria, L. (2012) Seed dispersal in changing landscapes. *Biological Conservation* 146, 1–13.

Meyer, S.T., Leal, I.R., Tabarelli, M. and Wirth, R. (2011) Performance and fate of tree seedlings on and around nests of the leaf-cutting ant *Atta cephalotes*: ecological filters in a fragmented forest. *Austral Ecology* 36, 779–790.

Moran, C., Catterall, C.P., Green, R.J. and Olsen, M.F. (2004) Functional variation among frugivorous birds: implications for rainforest seed dispersal in a fragmented subtropical landscape. *Oecologia* 141, 584–595.

Moran, C., Catterall, C. and Kanowski, J. (2009) Reduced dispersal of native plant species as a consequence of the reduced abundance of frugivore species in fragmented rainforest. *Biological Conservation* 142, 541–552.

Morris, R.J. (2010) Anthropogenic impacts on tropical forest biodiversity: a network structure and ecosystem functioning perspective. *Philosophical Transactions of The Royal Society B: Biological Sciences* 365, 3709–3718.

Mosisch, T.D., Bunn, S.E. and Davies, P.M. (2001) The relative importance of shading and nutrients on algal production in subtropical streams. *Freshwater Biology* 46, 1269–1278.
Mouquet, N., Grave, D., Massol, F. and Calcagno, V. (2013) Extending the concept of keystone species to communities and ecosystems. *Ecology Letters* 16, 1–8.
Nascimento, H.E.M. and Laurance, W.F. (2004) Biomass dynamics in Amazonian forest fragments. *Ecological Applications* 14, S127–S138.
Neuschulz, E.L., Botzat, A. and Farwig, N. (2011) Effects of forest modification on bird community composition and seed removal in a heterogeneous landscape in South Africa. *Oikos* 120, 1371–1379.
Ostfield, R.S. and LoGiudice, K. (2003) Community disassemby, biodiversity loss, and the erosion of an ecosystem service. *Ecology* 84, 1421–1427.
Pascoal, C., Cássio, F., Marcotegui, A., Sanz, B. and Gomes, P. (2005) Role of fungi, bacteria, and invertebrates in leaf litter breakdown in a polluted river. *Journal of the North American Benthological Society* 24, 784–797.
Pastor, J. and Naiman, R.J. (1992) Selective foraging and ecosystem processes in boreal forests. *American Naturalist* 139, 690–705.
Paul, M.J., Meyer, J.L. and Couch, C.A. (2006) Leaf breakdown in streams differing in catchment land use. *Freshwater Biology* 51, 1684–1695.
Peh, K.S.-H. and Lewis, S.L. (2012) Conservation implications of recent advances in biodiversity-functioning research. *Biological Conservation* 151, 26–31.
Phillips, O.L., Hall, P., Gentry, A.H., Sawyer, S.A. and Vásquez, R. (1994) Dynamics and species richness of tropical rain-forests. *Proceedings of the National Academy of Sciences USA* 91, 2805–2809.
Pillai, P., Gonzalez, A. and Loreau, M. (2011) Metacommunity theory explains the emergence of food web complexity. *Proceedings of the National Academy of Sciences USA* 108, 19293–19298.
Pusey, B.J. and Arthington, A.H. (2003) Importance of the riparian zone to the conservation and management of freshwater fish: a review. *Marine and Freshwater Research* 54, 1–16.
Rantalainen, M.-L., Fritze, H., Hhaimi, J., Pennanen, T. and Setälä, H. (2005) Species richness and food web structure of soil decomposer community as affected by the size of habitat fragment and habitat corridors. *Global Change Biology* 11, 1614–1627.
Reich, P.B., Tilman, D., Naeem, S., Ellsworth, D.S., Knops, J., Craine, J., Wedin, D. and Trost, J. (2004) Species and functional group diversity independently influence biomass accumulation and its response to CO_2 and N. *Proceedings of the National Academy of Sciences USA* 101, 10101–10106.
Restrepo, C. and Vargas, A. (1999) Seeds and seedlings of two neotropical montane understory shrubs respond differently to anthropogenic edges and treefall gaps. *Oecologia* 119, 419–426.
Ricketts, T.H. (2001) The matrix matters: effective isolation in fragmented landscapes. *American Naturalist* 158, 87–99.
Ricketts, T.H., Daily, G.C., Ehrlich, P.R. and Michener, C.D. (2004) Economic value of tropical forest to coffee production. *Proceedings of the National Academy of Sciences USA* 101, 12579–12582.
Riutta, T., Slade, E.M., Bebber, D.P., Taylor, M.E., Malhi, Y., Riordan, P., Macdonald, D.W. and Morecroft, M.D. (2012) Experimental evidence for the interacting effects of forest edge, moisture and soil macrofauna on leaf litter decomposition. *Soil Biology and Biochemistry* 49, 124–131.
Rosenlew, H. and Roslin, T. (2008) Habitat fragmentation and the functional efficiency of temperate dung beetles. *Oikos* 117, 1659–1666.
Rubinstein, A. and Vasconcelos, H.L. (2005) Leaf-litter decomposition in Amazonian forest fragments. *Journal of Tropical Ecology* 21, 699–702.
Schleuning, M., Farwig, N., Peters, M.K., Bergsdorf, T., Bleher, B., Brandl, R., Dalitz, H., Fischer, G., Freund, W., Gikungu, M.W., Hagen, M., Garcia, F.H. *et al.* (2011) Forest fragmentation and selective logging have inconsistent effects on multiple animal-mediated ecosystem processes in a tropical forest. *PLoS One* 6, e27785.
Schmitz, O.J., Hambäck, P.A. and Beckerman, A.P. (2000) Trophic cascades in terrestrial systems: a review of the effects of carnivore removal on plants. *American Naturalist* 155, 141–153.
Schnitzer, S.A. and Bongers, F. (2011) Increasing liana abundance and biomass in tropical forests: emerging patterns and putative mechanisms. *Ecology Letters* 14, 397–406.
Schüepp, C., Herrmann, J.D., Herzog, F. and Schmidt-Entling, M.H. (2011) Differential effects of habitat isolation and landscape composition on wasps, bees, and their enemies. *Oecologia* 165, 713–721.
Slade, E.M., Mann, D.J., Villanueva, J.F. and Lewis, O.T. (2007) Experimental evidence for the effects of dung beetle functional group richness and composition on ecosystem function in a tropical forest. *Journal of Animal Ecology* 76, 1094–1104.
Slade, E.M., Mann, D.J. and Owen, T.L. (2011) Biodiversity and ecosystem function of tropical forest dung beetles under contrasting logging regimes. *Biological Conservation* 144, 166–174.

Sodhi, N.S., Koh, L.P., Clements, R., Wanger, T.C., Hill, J.K., Hamer, K.C., Clough, Y., Tscharntke, T., Posa, M.R.C. and Lee, T.M. (2010) Conserving Southeast Asian forest biodiversity in human-modified landscapes. *Biological Conservation* 143, 2375–2384.

Sole, R.V. and Montoya, J.M. (2006) Ecological network meltdown from habitat loss and fragmentation. In: Pascual, M. and Dunnem, J.A. (eds) *Ecological Networks: Linking Structure to Dynamics in Food Webs*. Oxford University Press, New York, pp. 305–324.

Sork, V. (1983) Distribution of pignut hickory (*Carya glabra*) along a forest to edge transect, and factors affecting seedling recruitment. *Bulletin of the Torrey Botanical Club* 110, 494–506.

Spiegel, O. and Nathan, R. (2007) Incorporating dispersal distance into the disperser effectiveness framework: frugivorous birds provide complementary dispersal to plants in a patchy environment. *Ecology Letters* 10, 718–728.

Tallmon, D.A., Jules, E.S., Radke, N.J. and Mills, L.S. (2003) Of mice and men and trillium: cascading effects of forest fragmentation. *Ecological Applications* 13, 1193–1203.

Tanentzap, A.J. and Coomes, D.A. (2012) Carbon storage in terrestrial ecosystems: do browsing and grazing herbivores matter? *Biological Reviews* 87, 72–94.

Terborgh, J., Lopez, L., Nunez, P., Rao, M., Shahabuddin, G., Orihuela, G., Riveros, M., Ascanio, R., Adler, G.H., Lambert, T.D. and Balbas, L. (2001) Ecological meltdown in predator-free forest fragments. *Science* 294, 1923–1926.

Tewksbury, J.J., Levey, D.J., Haddad, N.M., Sargent, S., Orrock, J.L., Weldon, A., Danielson, B.J., Brinkerhoff, J., Damschen, E.I. and Townsend, P. (2002) Corridors affect plants, animals, and their interactions in fragmented landscapes. *Proceedings of the National Academy of Sciences USA* 99, 12923–12962.

Tewksbury, J.J., Garner, L., Garner, S., Lloyd, J.D., Saab, V. and Martin, T.E. (2006) Tests of landscape influence: nest predation and brood parasitism in fragmented ecosystems. *Ecology* 87, 759–768.

Tilman, D. (1999a) Diversity and production in European grasslands. *Science* 286, 1099–1100.

Tilman, D. (1999b) The ecological consequences of changes in biodiversity: a search for general principles. *Ecology* 80, 1455–1474.

Tilman, D., Reich, P.B., Knops, J., Wedin, D., Mielke, T. and Lehman, C.L. (2001) Diversity and productivity in a long-term grassland experiment. *Science* 294, 843–845.

Trumbo, S.T. and Bloch, P.L. (2001) Habitat fragmentation and burying beetle abundance and success. *Journal of Insect Conservation* 4, 245–252.

Turner, M.G. (2005) Landscape ecology: what is the state of the science? *Annual Review of Ecology, Evolution, and Systematics* 36, 319–344.

Tylianakis, J.M., Didham, R.K., Bascompte, J. and Wardle, D.A. (2008) Global change and species interactions in terrestrial ecosystems. *Ecology Letters* 11, 1351–1363.

Valladares, G., Cagnolo, L. and Salvo, A. (2012) Forest fragmentation leads to food web contraction. *Oikos* 121, 299–305.

Vasconcelos, H.L. and Laurance, W.F. (2005) Influence of habitat, litter type, and soil invertebrates on leaf-litter decomposition in a fragmented Amazonian landscape. *Oecologia* 144, 456–462.

Vilà, M., Vayreda, J., Comas, L., Ibáñez, J.J., Mata, T. and Obón, B. (2007) Species richness and wood production: a positive association in Mediterranean forests. *Ecology Letters* 10, 241–250.

Walker, B. (1995) Conserving biological diversity through ecosystem resilience. *Conservation Biology* 9, 747–752.

Weathers, K.C., Cadenasso, M.L. and Pickett, S.T.A. (2001). Forest edges as nutrient and pollutant concentrators: potential synergisms between fragmentation, forest canopies, and the atmosphere. *Conservation Biology* 15, 1506–1514.

Williams-Linera, G. (1990) Vegetation structure and environmental conditions of forest edges in Panama. *Journal of Ecology* 78, 356–373.

Wirth, R., Meyer, S.T., Leal, I.R. and Tabarelli, M. (2008) Plant herbivore interactions at the forest edge. *Progress in Botany* 69, 423–448.

Young, A. and Mitchell, N. (1994) Microclimate and vegetation edge effects in a fragmented podocarp–broadleaf forest in New Zealand. *Biological Conservation* 67, 63–72.

Young, R.G., Matthaei, C.D. and Townsend, C.R. (2008) Organic matter breakdown and ecosystem metabolism: functional indicators for assessing river ecosystem health. *Journal of the North American Benthological Society* 27, 605–625.

Zavaleta, E., Pasari, J., Moore, J., Hernández, D., Suttle, K.B. and Wilmers, C.C. (2009) Ecosystem responses to community disassembly. *Annals of the New York Academy of Sciences* 1162, 311–333.

9 Beyond the Mirror: Tropical Forest Fragmentation and its Impact on Rural Livelihoods

Claude A. Garcia[1,2]* and Laurène Feintrenie[2]
[1]*ETH Zurich, Switzerland;* [2]*CIRAD, Montpellier, France*

Abstract

Forest fragmentation is overwhelmingly the result of changes in land use and the development of infrastructures. It can be considered as the spatial manifestation of deforestation and degradation in a landscape. Forest fragmentation thus results in the same loss of ecosystem services as these other forms of land-cover change, but how does it affect livelihoods? We propose here an analysis of the livelihood impacts of tropical forest fragmentation, which will begin with a better definition of the spatial processes involved in fragmentation. As it is difficult to analyse the impacts of fragmentation without consideration of the associated impacts of the causes – or drivers – of forest fragmentation, we will analyse them in detail. Fragmentation results from a range of drivers that are categorized into five groups: (i) economic factors; (ii) governance factors; (iii) demographic factors; (iv) technological factors; and (v) cultural factors. Economic factors are prevalent in decisions, with forest conservation often being a luxury for wealthy people. Stakeholders who gain direct economic returns from deforestation – from logging, farming and facilitated access to markets – will also benefit globally from forest fragmentation, the benefits from these drivers by far outweighing the loss of ecosystem services. However, forest dwellers and isolated forest communities will often have to choose between 'modern' life or the cultural specificities that shaped their identity. Throughout the chapter, the forest-transition curve is used as a heuristic framework to understand the relationships between human development and the loss and fragmentation of forest cover.

Understanding Fragmentation

The concept of fragmentation

A fragment is a broken piece detached from a larger body and thus fragmentation is the process of breaking up an entity into smaller pieces. As a concept, fragmentation is used in anthropology and sociology to denote the divides appearing between groups (Easterly and Levine, 1997; Jenks, 2004), and in history and political sciences to describe the collapse of larger institutions into smaller ones (Fukuyama, 2007; Morris, 2010). In ecology, Preston and then MacArthur and Wilson suggested that species assemblages in mainland habitats broken into smaller isolated pieces may behave similarly to true islands (Preston, 1962; MacArthur and Wilson, 1967). Since then, fragmentation, and particularly habitat

*claude.garcia@usys.ethz.ch

fragmentation, has become a major topic for scientists, conservationists and managers, with thousands of published references (Collinge, 2009). However, different authors have used different definitions and different metrics (Fahrig, 2003). Fragmentation has come to represent all kinds of human-driven ecosystem changes and patterns, and this ambiguity has diminished the usefulness of the concept (Haila, 2002; Lindenmayer and Fischer, 2007). The concept generally refers to the process by which a 'habitat is transformed into a number of smaller patches of smaller total area, isolated from each other by a matrix of habitats unlike the original' (Wilcove *et al.*, 1986, p. 237).

One of the problems with this definition, and indeed with most of the work done on habitat fragmentation, is that it conflates habitat loss ('of smaller total area', in the definition above) with the breaking apart of habitats and their subsequent isolation, making it difficult to tease out the differential effects of both processes.

Our objective in this chapter is to review the impact of forest fragmentation on the livelihood of forest dwellers and other forest-dependent communities, with an emphasis on tropical socio-ecosystems. This is unusual as most authors working on forest fragmentation have considered the impact of human activities on the ecosystems and on biodiversity (Tole, 2006; Fischer and Lindenmayer, 2007) but seldom have addressed the feedback loop by which societies react to the new landscapes they have created.

Following Fahrig (2003), we try to focus here on the effects of fragmentation per se; that is, changes in habitat configuration, independently from habitat loss, avoiding thus writing one more review on deforestation and its impact on livelihood. However, deforestation, degradation and fragmentation are so intimately linked that this distinction, theoretically sound as it may be, may hinder more than help our argument (Didham *et al.*, 2012), so we will take some liberties with it.

Fragmentation studies are sensitive to the number of classes used to construct the landscape typologies (Turner, 1989; Rutledge, 2003). For the purpose of our discussion, we will restrict ourselves to forest as the habitat in the definition by Wilcove *et al.* (1986), and agricultural land and infrastructures as the matrix that surrounds the forest fragments. With such a binary and artificial boundary, we will miss the important topic of the impact of forest degradation, defined as the reduction in the capacity of a forest to produce ecosystem services (Thompson *et al.*, 2013).

In recent years the research community has progressively shifted its focus from fragments to landscapes (Kupfer *et al.*, 2006). Fischer and Lindenmayer (2007) proposed a holistic framework to reconcile the multiplicity of approaches of fragmentation, from species-oriented to pattern-oriented insights. But even that fails to consider the links between the landscapes and the societies that shape them in terms other than disturbances and threats.

The field has not been left vacant though, as other disciplines – rural development, economy and health, to name but a few – have addressed the topic of the livelihood impacts of land-use change and deforestation. However, authors from these disciplines are unlikely to frame their problems as one of forest fragmentation (but see Allan *et al.*, 2003).

We will first address the impacts that fragmentation of forests have on the access to forest resources and on the livelihood strategies of forest dwellers and forest-dependent communities. We will then shift our focus from the fragments to the matrix, as, if the objective is to conserve biodiversity in the fragments, then what matters is how the matrix is managed (Crome, 1997). In doing so, we will demonstrate that the process of fragmentation is a problem for forest managers and conservationists but not necessarily for a forest dweller or an agriculturist in a deforestation front (Fig. 9.1).

Scales and processes

A further difficulty that fragmentation studies need to deal with is the different scales at which processes unfold. Fragmentation is by essence a landscape-level process and cannot be measured at the patch level (Fahrig, 2003), and, as deforestation and fragmentation happen, the scale itself at which forests dominate a landscape will change (Wickham *et al.*, 2007).

The heuristic framework of the forest-transition curve sheds light on the relationships between human development and the evolution of forest cover (Mather, 1992, 2007; Rudel *et al.*, 2005). The theory has been developed at the scale of a nation, but similar trajectories have been described to extend across entire regions, suggesting common underlying

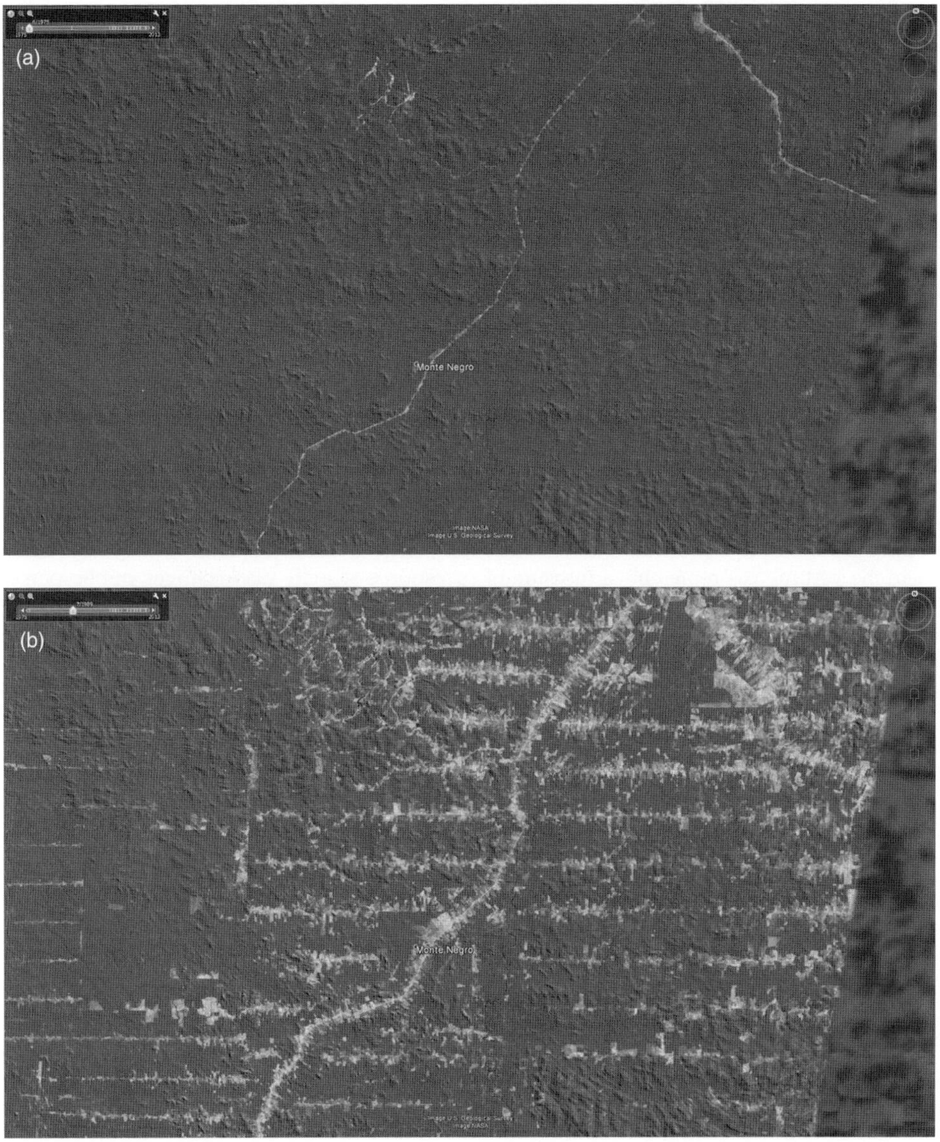

Fig. 9.1. Forest fragmentation and development in the state of Rondônia, Brazil. Landsat images for 1975 (a), 1989 (b) and 2001 (c, see next page). Agricultural land and urban areas appear light coloured and forests are darker coloured. Infrastructure and road development have paved the way for the conversion of forest to agriculture, creating a typical fishbone pattern. In this case, fragmentation has preceded deforestation and degradation. However, the fragmentation gradient correlates with a gradient of connection to market and services. The point of view of a rural developer is the inverted image of the one of the forester (from Tejaswi and Holmgren, 2007). Sources: Google, US Geological Survey and NASA.

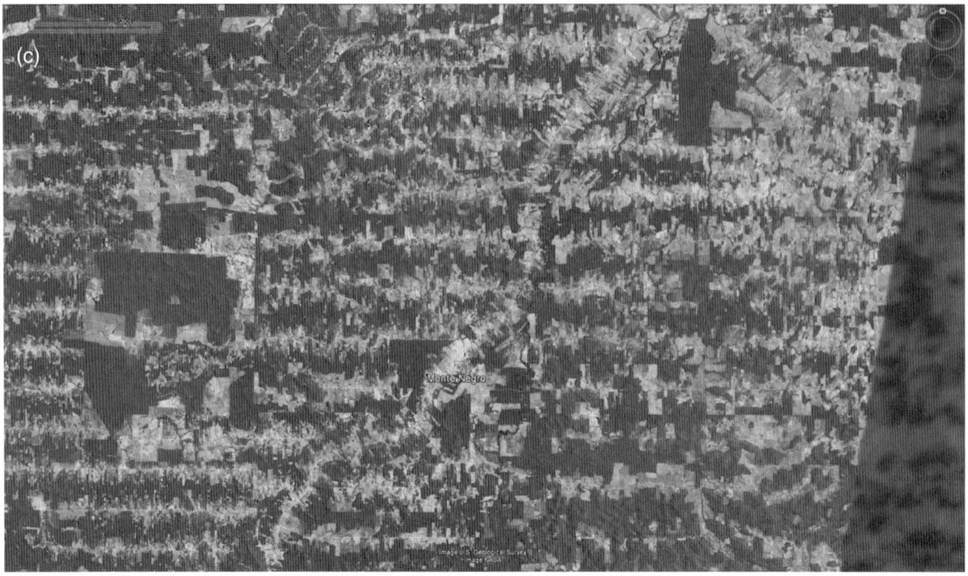

Fig. 9.1. Continued.

historical, social and environmental conditions (Rudel *et al.*, 2005). This is illustrated in Fig. 9.2, where the vertical axis represents a measure of forest cover (or biomass) and the horizontal axis is loosely defined as development or demographic pressure on land (Mather and Needle, 1998). It could represent time, albeit in a non-linear fashion, with some phases, the initial one for example, lasting for centuries whereas others can happen within a lifetime or even faster. The timescale can thus be contracted or expanded for the purpose of intelligibility (Fig. 9.2).

The forest-transition curve has two phases, reflecting the loss and subsequent recovery of forest as a society develops and its population grows. During the first phase, logging operations and agricultural clearings reduce forest cover below its theoretical dynamic equilibrium level. Increased pressure on land and the demand for forest resources and energy stimulates the expansion in arable land and associated infrastructures, gradually reducing the forest cover until it reaches a minimum. Deforestation will eventually stop and, as the society develops its infrastructure and economic activities, the change from contraction to expansion of its forest area is operated through the concentration of agricultural production in the most productive lands, leaving suboptimal areas available for reforestation through natural regeneration or planting. Forest and forest resource scarcity also changes public perceptions, and policies emerge that reinforce this upward trend in forest cover. These two different pathways to recovery – reduced pressure on forests and oriented plantation programmes in response to scarcity – have been described in European and Asian societies, respectively (Rudel *et al.*, 2005). The forests of the latest phase, a mix of remnants, regrowth and plantations, differ from the earlier ones, particularly in terms of biodiversity. However, in other aspects – carbon storage for example – they may well be comparable (Kauppi *et al.*, 2006). Recovery of the forest cover is not ineluctable though, and poverty traps can prevent the transition from happening (Rudel *et al.*, 2005).

The forest-transition curve is a theory of how deforestation happens globally. Transitions, when they happen, are neither deterministic nor do they follow a fixed pattern (Lambin *et al.*, 2003). They are 'possible development paths where the direction, size and speed can be influenced through policy and specific circumstances' (Martens and Rotmans, 2005, p. 1136). The drivers of forest transitions are multiple and interrelated, combining (i) underlying economic factors,

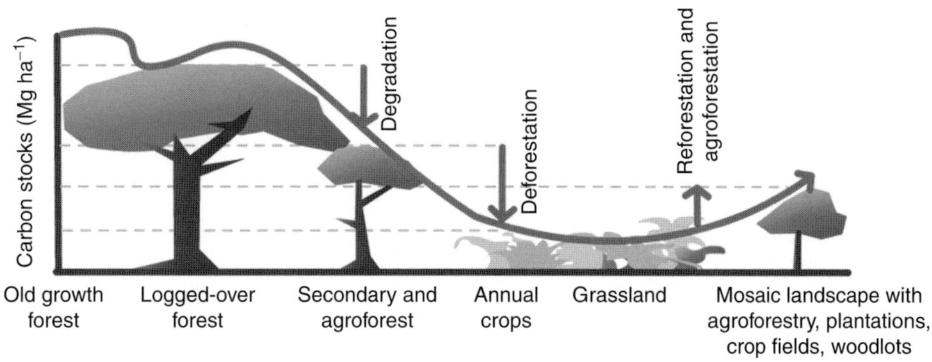

Fig. 9.2. The forest-transition curve. Forest cover experiences phases of shrinkage and expansion. Simultaneous with these changes in extent, the nature of the forest cover itself is modified, with old-growth forests progressively being replaced by secondary forests and later plantations and agroforestry mosaics. The horizontal axis loosely represents development and increased population pressures. (From CRP6, 2011.)

institutions and national policies, and (ii) direct agricultural expansion, wood extraction and infrastructure extension (Geist and Lambin, 2002). However, it is possible to identify a limited set of pathways, allowing the identification of general principles driving land-use changes (Lambin *et al.*, 2003) and thus forest fragmentation.

McIntyre and Hobbs (1999) described a similar pattern with their conceptual model of landscape change. Starting from a landscape reputedly (i) intact, in which the forest cover was little or unmodified, they proposed a succession of stages described as (ii) variegated, where the landscape is dominated by the original vegetation, but with gradients of habitat modification; (iii) fragmented, in which only fragments of the original vegetation remain, in a matrix of other land uses; and (iv) relict, where less than 10% of the original cover remains, in a starkly contrasting matrix. The authors recognized that different stages along this pathway pose different kinds of challenges for conservation management (McIntyre and Hobbs, 1999). What they proposed is in essence the local, spatially explicit translation of the global forest-transition theory, applied to a landscape. Similarly, Forman (1995) identified five different spatial processes that are related to landscape change: (i) perforation; (ii) dissection; (iii) subdivision (or fragmentation); (iv) shrinkage; and (v) attrition (Fig. 9.3). The word 'fragmentation' has been used to represent some or all five of these distinct spatial processes and the patterns they create, yet the drivers and impacts of each of these processes differ. In addition, as with McIntyre and Hobbs (1999), these five processes follow a temporal pattern that maps out the first phase of the forest transition – perforation and dissection when the forest starts to be modified, fragmentation when the pace of transformation hastens, and shrinkage and attrition when the forest cover reaches the lowest levels (Forman, 1995).

Fragmentation thus appears as the local, spatially explicit translation of the global process of deforestation and land-use change as theorized by the forest-transition model. This understanding will help us analyse the drivers of landscape change (e.g. population growth, market access) and the contexts under which they operate (e.g. law enforcement, mode of governance) as their combination determines the impacts on forest cover and therefore the spatial structures that deforestation creates (Sassen *et al.*, 2013).

Patterns and drivers

The processes of fragmentation as defined by Wilcove *et al.* (1986) imply four distinct effects on the landscape: (i) a reduction in total forest cover; (ii) an increase in the number of forest patches; (iii) a decrease in mean size of forest patches; and finally (iv) an increase in the total amount of forest/matrix

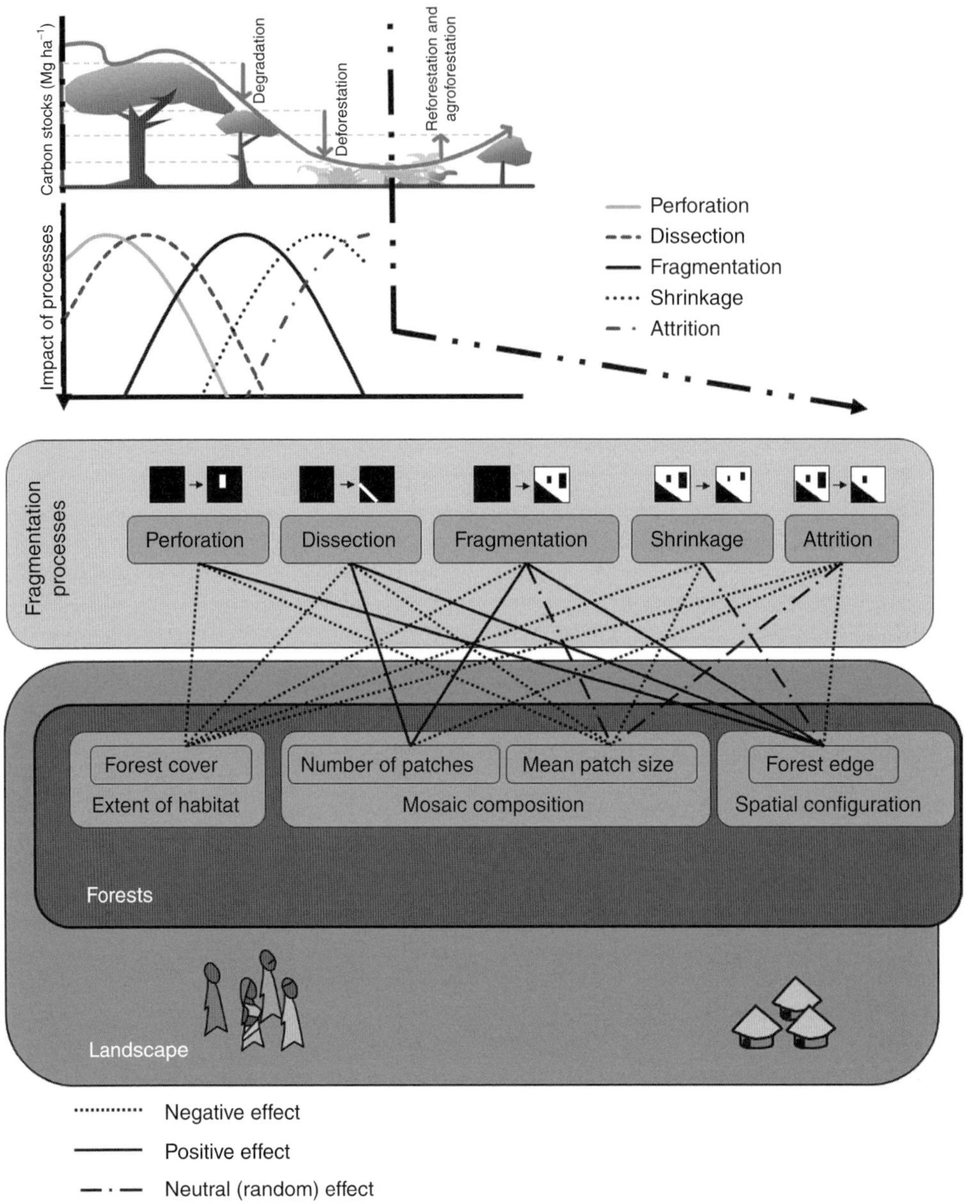

Fig. 9.3. Heuristic framework on fragmentation processes, their links with the forest-transition curve and their impact on the landscape mosaic. As deforestation happens, the five processes of fragmentation – perforation, dissection, fragmentation, shrinkage and attrition – will successively take precedence in shaping the landscape mosaic. Applied to forests, these processes will have positive, negative or neutral impacts on the extent of habitat, the mosaic composition and the spatial configuration. These changes will in turn have an impact on the livelihoods of the people in the landscape, beyond the processes happening inside the forest fragments. (Adapted from Forman, 1995, and CRP6, 2011.)

edge (Fahrig, 2003; Rutledge, 2003). In addition, fragmentation is a non-random process (Dale and Pearson, 1997), and the ensuing spatial distribution and shape of the fragments will also have an impact on the system. All these different patterns of ecosystem change are interrelated and difficult, if not impossible, to analyse independently empirically.

It is, however, possible to link the processes of fragmentation to the patterns that they create (Fig. 9.3). All five processes (perforation, dissection, fragmentation, shrinkage and attrition) will reduce forest cover in the landscape. The number of forest fragments (patches) increases with dissection and fragmentation but is reduced by attrition, whereas perforation and shrinkage have no impact. Mean patch size tends to diminish with perforation, dissection and shrinkage. Fragmentation and attrition can increase or decrease mean patch size depending on which fragment they affect. Finally, the length of forest edge will increase with perforation, dissection and fragmentation as new edges are created, and diminish with attrition when fragments disappear. Shrinkage is neutral, as it can reduce edge or increase it, depending on the spatial configurations. The complexity of these interactions, the multitude of parameters and the gradation in time explain why it is difficult to describe simple linear functions of the impact of fragmentation on any given variable, be it ecological or social.

As fragmentation is the spatially explicit form that deforestation takes in a landscape, the same causes that drive deforestation will also drive fragmentation in tropical landscapes. Deforestation, degradation and fragmentation generally result from a combination of multiple causes, both direct, linked to the opening of the forest, and indirect, comprising the background forces that make forest clearing the rational choice for the stakeholders involved (Kanninen *et al.*, 2007). Despite the diversity of possible causes and enabling factors, there is a small set of patterns that emerges from the interactions between these drivers, and it is possible to identify similar regional and historical pathways (Lambin *et al.*, 2003). Of the three major sources of deforestation – agricultural expansion, infrastructure development, and wood extraction (Geist and Lambin, 2002) – agricultural expansion is by far the most prevalent land-use change associated with forest cover loss. Shifting cultivation is generally restricted to the initial phases of fragmentation (perforation), but the development of the road network (roads, railroads) will allow penetration into the forest (dissection), creating the familiar fishbone patterns seen in Fig. 9.1 and paving the way for other drivers to take effect (Kanninen *et al.*, 2007). Large infrastructure developments, including dams and mining operations, can also kick-start the forest transition (perforation and dissection). Settled agriculture, cattle ranching and colonization (transmigration and resettlements) will permanently open large tracts of forest (fragmentation). The development of settlements will put additional pressure on the remaining forest fragments (shrinkage and attrition). The requirements for timber and poles, fuel wood and charcoal will contribute to degrade the forest remnants, in turn increasing the probability that they disappear, as they lose value as forest (shrinkage).

The underlying drivers of deforestation, degradation and fragmentation typically unfold at the regional level, unlike the direct causes that are generally local and play at the farm, household or community level (Lambin *et al.*, 2003). These underlying drivers are often exogenous to the social and ecological system undergoing change, and out of the control of the agents involved in deforestation and fragmentation. They are categorized into five clusters: (i) economic factors; (ii) governance factors; (iii) demographic factors; (iv) technological factors; and (v) cultural factors, acting in synergy to enable the proximate drivers to unfold (Geist and Lambin, 2002). Of these five, economic factors are the most prevalent. The root cause of forest conversion and fragmentation is the higher profitability of agriculture (Wunder and Verbist, 2003). Economic development and transitions to cash-based economies through the connection to markets are a strong explanatory factor of deforestation. Higher demand for agricultural products, currency devaluations, structural adjustments, trade policies protecting extensive agriculture and a subsidized timber sector have all been documented to contribute to deforestation and its spatial impact, forest fragmentation (Kaimowitz and Angelsen, 1998; Angelsen and Wunder, 2003). Governance factors also play a major role in explaining deforestation. The nature of tenure rights (Agrawal and Ostrom, 2001), the strength and clarity of the institutions, rights and responsibilities (Ostrom, 1990, 1999), transparency (Barr, 2001; Casson and Obidzinski, 2002) and law and its enforcement (Colchester, 2006) all define the context that enable the primary drivers to unfold

(Sassen *et al.*, 2013). Demographic factors are commonly held as the primary root cause for deforestation, with higher populations increasing the demand for food and space and ensuring there is an available labour force to cut down the trees. This is, however, a misconception. Strikingly, only in-migration of settlers into 'empty' forest areas appears to have a significant impact on deforestation, mainly in Africa and Latin America. Population increases through high fertility rates only contribute to explain deforestation in a minority of cases, and always combined with other factors (Geist and Lambin, 2002). In addition, the bulk of the land-use change is caused by industrialized agriculture rather than by the more labour-demanding smallholder agriculture (Kanninen *et al.*, 2007). Technological factors will affect deforestation rates, but they can play both ways. New technologies can improve the profitability of alternative land uses, increasing the pressure on forests, leading to the *intensification* and *expansion* outcome described in the literature (Angelsen and Kaimowitz, 2001). They can also provide new opportunities and alternative livelihoods, diverting people and resources from the forest and thereby reducing pressure (Chomitz *et al.*, 2007; Kanninen *et al.*, 2007). Finally, cultural factors such as the attitudes towards forests, a 'frontier' mentality or a rent-seeking behaviour will also determine the choices of individuals engaged in deforestation (Bhagwat *et al.*, 2005b; Garcia and Pascal, 2006; Geist *et al.*, 2006).

All these proximate and underlying drivers of deforestation, degradation and forest fragmentation have been documented to interact in multiple and complex ways, making it impossible to identify direct causal links. While there are some invariants, like the development of markets and the expansion of permanent crop land, most explanations are regional and historically bounded. While large-scale, global and national drivers (policies and markets) set the backdrop and enable deforestation to happen, fragmentation, its local, spatially explicit form, is driven mainly by agricultural expansion and road development (Geist and Lambin, 2002). As such, no universal policy can be recommended to reverse the process, and solutions will need to be tailored locally to the specific problem they address.

Fragmentation and Livelihoods

As described in Fig. 9.2, fragmentation in a tropical forest landscape will thus modify the extent and disposition of forests in the landscape. To get a clearer picture of the impact of fragmentation on livelihoods, we will look in more detail at how changes in these three landscape mosaic parameters (extent of forest cover, mosaic composition, and spatial configuration) affect local communities and their livelihoods. Here, we follow the definition of livelihood as comprising the five different assets or capitals (natural, physical, human, financial and social) that stakeholders can tap into, their access to these, which is determined by the institutional and governance framework they live in, and the practices and strategies they develop to secure their quality of life using these assets (Ellis, 2000).

Winners and losers of forest cover loss

The magnitudes of the impacts of forest loss and forest fragmentation differ. On biodiversity, the impact of forest cover loss far outweighs those of fragmentation per se, to the point that it makes it difficult to detect the latter empirically (Fahrig, 2003). On livelihoods, the same trend is expected. However, in addressing the impacts of the changes, we also need to consider the impacts of the direct drivers (agriculture expansion, infrastructure development and wood extraction) behind the landscape changes, lest we provide a biased account of the processes.

The links between forest cover loss and poverty in developing countries and the impact of deforestation on local livelihoods are well documented, and a detailed overview would fall outside the purpose of this chapter (for a review, see Sunderlin *et al.*, 2005). We simply stress that a reduction in forest cover results in positive and negative livelihood outcomes, depending on the perspectives and livelihood strategies of the stakeholders involved (Table 9.1.) (Contreras-Hermosilla, 2000).

Table 9.1. The winners and losers of forest cover loss. (Adapted from Contreras-Hermosilla, 2000.)

Stakeholders	Impact of forest loss and its direct drivers	References
Forest dwellers	Loss of spiritual values and identity; loss of traditional knowledge; loss of forest resources and hunting grounds; improved access to markets; improved access to services	Kushalappa and Bhagwat, 2001; Pritchard, 2004; Godoy *et al.*, 2005; Levang *et al.*, 2007
Farmers and cultivators	Changing environmental conditions, particularly water; reduced pressure from wildlife; increased agricultural area; improved access to markets	Williams, 2003; Sunderlin *et al.*, 2005; Scanlon *et al.*, 2007; Feintrenie *et al.*, 2010; Garcia *et al.*, 2010
Urban dwellers	Loss of forest resources, e.g. bushmeat, charcoal; loss of recreational amenities and recreational space; loss of ecosystem services particularly cheap clean water supply; more housing space; increased flow of agricultural products; growing concerns over loss of biodiversity	Kahn, 2000; MacDonald and Rudel, 2005; Brauman *et al.*, 2007; Nasi *et al.*, 2008; Schure *et al.*, 2011
Logging industry and supply chain	Increased short term job opportunities and profits; loss of long-term prospects; better market access	Medjibe and Putz, 2012; Gourlet-Fleury *et al.*, 2013
Mining and other industries	Possible negative public image; improved access to resources (non-forest)	Megevand, 2013
National governments and decision makers	Reduced political pressure over landlessness; increased international pressure to halt deforestation; loss of prospects of revenue from timber industry; increased revenue from other sectors (mining, agriculture)	Fearnside, 2005; Rudel *et al.*, 2005; Lambin and Meyfroidt, 2010

This table is not exhaustive but serves to highlight the multiplicity of perspectives, and the trade-offs between stakeholders affected by the loss of forest cover and the three majors direct drivers: agricultural expansion, wood extraction and infrastructure development. This table focuses specifically on the local impacts, more relevant to the problem of fragmentation.

With the exception of the forest-dependent communities, most stakeholders stand to gain from the conversion of forests to agricultural land and the development of infrastructures. This is not a surprise and explains the historical trends of agriculture expansion and development (Williams, 2003). We simply outline that what is described as a global trend also holds when we look at the local scale, and that win–lose scenarios where livelihoods improve while forest cover decreases are the norm (Sunderlin *et al.*, 2005). For most forest-dependent communities, however, the prospects are dire. Seldom do they have the luxury to benefit from improved access to services (health and education) without the concomitant forceful integration into the market economy and the loss of their traditions, knowledge and identity (Levang *et al.*, 2007).

Many of the changes outlined here will in turn feed back on the drivers of deforestation and fragmentation, slowing and dampening the process as the negative impacts accumulate (Lambin and Meyfroidt, 2010). They can, however, also create reinforcing loops, for example when exposure to market and increased income leads to increased pressure on forest, even in the more traditional and remote forest-dwelling communities (Godoy *et al.*, 2009).

The livelihood value of small isolated forest fragments

Forests contribute to the subsistence and the income of local communities, in both humid and dry environments (Byron and Arnold,

1999; Shackleton *et al.*, 2007). But how does this relationship change with fragmentation? The conservation value of tropical rainforest fragments was established by Turner and Corlett (1996). For biologists and conservationists, the answer is clear: 'Fragments are better than nothing' (Turner and Corlett, 1996, p. 331). Here, we reverse the question, exploring their livelihood value, and the impacts that dissection and fragmentation will have on the people living in mosaics with forest fragments. How important are these fragments to local people? This question is not commonly addressed in the literature, with the notable exception of the sacred forests (Chandrakanth and Nagaraja, 1997).

Urech *et al.* (2012) explored this on the east coast of Madagascar. Their results show that interest in the preservation of forest fragments varies according to two variables: distance to forest and household income. As distance to the forest grew, so did awareness of the importance of fragments as source of wood and non-timber forest products. Similarly, the interest for the conservation of the fragments grew with the wealth of the household, not with the relative importance of the forest in the household income. Perhaps surprisingly and against commonly held visions, it is not thus dependence on forest resources that leads to appreciation of the forest and the fragments, but scarcity, which comes with increased distance from the forest, and wealth, which frees the household from the need to access the resource for its subsistence and allows it to consider conservation as a possibility (Urech *et al.*, 2012). A similar result was found in the Western Ghats of India, where the farmers that would maintain forest fragments in their private lands were the ones without the capital to transform them into profitable coffee plantations, or the ones wealthy enough not to depend on the income that these could generate and had a genuine concern for biodiversity conservation. For the others, 'more than help from forests we have problems from it, like elephants' (Garcia *et al.*, 2007, p. 209). Such examples highlight that the links between forest fragments and local livelihoods are more complex and ambivalent than the literature tends to convey, and it is not simply proximity or dependence that explains the maintenance of forest fragments in a landscape, but rather: (i) the lack of capacity that comes from poverty; (ii) the perception of scarcity that only emerges in the final stages of the landscape conversion; and (iii) the willingness to protect that only comes with affluence.

The case of sacred forests deserves particular attention. India has probably the highest concentration of sacred forests (Ormsby and Bhagwat, 2010) and an equally high concentration of scholars interested in these social and ecological systems. As of November 2013, a Google Scholar search on 'sacred forest' yielded 260,000 hits compared with 95,900 for a search with 'sacred forest India' and only 23,600 for 'sacred forest Ethiopia'. The sacred forests in India and the traditions attached to them are as diverse as the country itself, and these traditions are well documented and reported in the literature (see, for example, Gadgil and Vartak, 1975; Ramakrishnan *et al.*, 1998; Malhotra *et al.*, 2001; Bhagwat *et al.*, 2005a; Ormsby and Bhagwat, 2010). In the south-east coast, sacred forests are the only dry evergreen forest remnants left in the landscape (Mani and Parthasarathy, 2005). Many authors report, however, that these fragments are threatened with conversion, and that attrition and shrinkage are also at play, even in the 'abodes of the gods' (Kushalappa and Bhagwat, 2001; Bhagwat and Rutte, 2006). Adopting a political ecology approach to the question, Freeman (1999) found a discrepancy between the environmentalist's models and the actual reasons for instituting and maintaining sacred forests at the local level in Kerala. Similarly, an analysis of the ecological status, the perceptions and the social role of the sacred forests of Kodagu, in the Western Ghats, highlighted that these fragments are not set aside. They are visited and managed and contribute directly to the material livelihoods of the local community. When shrinkage and attrition happen, it is generally the result of a neighbour pushing his fence inside the sacred forest. In addition, the more vocal advocates of their protection are often far removed from the day-to-day realities (Garcia and Pascal, 2006). Here also, distance seems to play a critical role in the appreciation of the forest fragments, even when it concerns the most intangible values.

By multiplying the number of fragments, and mechanically reducing the distance to the forest, it could be argued that fragmentation makes forest easier to access to the most marginalized elements of a community. Reduced to tiny forest patches, the fragments lose interest in the eyes of most stakeholders. They can afford to relinquish control over them as easily as they can dismiss them. But for some, even these tiny remnants can make a critical difference to their livelihoods. In a landscape that has undergone massive fragmentation, the only forests left are the forbidden ones, protected by the government and the forest stewards, the sacred ones, respected out of their symbolic value, and the invisible ones, those that are hidden or that nobody cares for except for the most humble.

Border fights, human wildlife conflict and the forest edge

The last tangible impact that fragmentation will have on the landscape is the increased length of forest–matrix interface. Perforation, dissection and fragmentation will all increase the ecotone. This will have consequences for the ecology of the forest, as well as for the livelihood of the people. By widening the zones of contact between natural habitats and agricultural and housing space, fragmentation increases the likelihood of interactions between humans, and the species they have domesticated, and the rich biodiversity of the tropical forest.

Large carnivores will switch from wild prey to livestock, with attacks happening more frequently in the deforestation front (Crawshaw, 2003). In the fragmented landscape of the southern Brazilian Amazon, jaguars and pumas responded to annual calving seasons and targeted medium-sized herds. As total forest cover declined and distance from the forest increased, attacks diminished, reflecting the loss of viable habitat range for the predators but also the nature of the radial development surrounding the cities in the study area (Michalski *et al.*, 2006). In Kibale National Park, in Uganda, crop losses to wildlife raids were greatest at the edge of the forest, and an individual's best strategy to avoid losses is to have a neighbour's field between their crops and the forest (Naughton-Treves, 1997). In the Western Ghats, the development of coffee cultivation over previously forested landscapes has increased the likelihood of encounters with elephants. This results in levels of conflicts and casualties three times higher than in the Trans Mara district in Kenya if reported by area (Bal *et al.*, 2011). The fact that no link was found between the composition of the agroforestry system and the frequency of elephant raids suggests that the elephants continue to use the paths that pre-existed prior to establishment of the plantations. In addition to the visible impacts, wildlife conflicts also entail 'hidden costs', including transaction costs occurring as a result of the encounter, and physical and psychological impacts on the affected population. These hidden costs negatively affect the well-being of the victims, yet rarely attract sufficient attention (Barua *et al.*, 2013). Such hidden costs also contribute to shape the attitude of the people towards nature and the forest, and taking them into account could help design strategies not to solve the problem of wildlife conflicts but to raise the tolerance level of the people living on the forest edge (Naughton-Treves, 1997; Jadhav and Barua, 2012).

Another facet of the increased interaction between forests and humans due to fragmentation is the emergence of new infectious diseases and the exchange of pathogens among wildlife, livestock and people. The outbreaks of severe acute respiratory syndrome (SARS), Ebola virus infection and the AIDS pandemic illustrate the importance of understanding the processes of zoonotic transmission, particularly between humans and non-human primates (Chapman *et al.*, 2005). A key element in the emergence of these new diseases is environmental change that allows increased contact with new host populations or even species (Daszak *et al.*, 2001). These changes can happen at the local scale, when humans and primates, for example, come into direct contact (hunting, crop raiding, research and tourism), at the regional scale, when habitats overlap (precisely the case when fragmentation occurs) and at the multiregional scale, when large-scale factors (for example, climate change) modify the patterns of transmission

(Chapman *et al.*, 2005). In forest fragments, results suggest that humans and non-human primates share parasites in Kibale, Uganda. Logging also increases the parasitic load of primates, suggesting that human activities associated with deforestation and fragmentation contribute to further increase the risk of pathogen exchange (Chapman *et al.*, 2005; Gillespie *et al.*, 2005).

Fire is another component of the forest–human interface with clear implications for livelihoods. Forest edges are dry and fire prone (Kapos, 1989), often in proximity to pastures that are regularly burned to control weeds and forest expansion. For example, fire occurrence in the eastern Brazilian Amazon is concentrated disproportionately near forest edges, suggesting a synergy between forest fragmentation and fire, with both processes mutually reinforcing each other (Cochrane and Laurance, 2002).

Not all interactions at the forest edge are detrimental to either the forests or the people. Trees and forests have positive effects on local property values when wildlife conflicts are not serious (MacDonald and Rudel, 2005), indicating that we tend to appreciate the proximity of the forest. Ecosystem services such as pest control depend on predators being able to stalk their prey in the agricultural field, and forest ecotones offer such vantage points.

What is the matrix?

We have analysed some of the livelihood impacts of fragmentations. From the moment the forest cover is open to the moment the last forest fragments vanish, the relative importance of the different processes we identified (perforation, dissection, fragmentation, shrinkage and attrition) will vary, as will their impact on the total forest cover, the composition of the landscape mosaic and the amount of forest boundary. As forest cover decreases, the spatial pattern of the remaining fragments gains importance. A question for the managers is whether these spatial patterns can be manipulated to compensate for the lost area (Rutledge, 2003). As forests are replaced by agricultural land and infrastructure, the nature of the matrix will play an ever-increasing role in dictating the ecological and social dynamics in the landscape (Kupfer *et al.*, 2006).

And as forests disappear from the landscape, so ends the story of fragmentation. A combination of growing environmental concerns, changing relative values between forests and non-forests, and development of other sectors of activity can eventually halt and even reverse deforestation and fragmentation. As farmers retire without successors, the valleys close up, woods expand and fragments merge. Plantations and agroforestry systems bring back trees in the landscape. These are not the same forests, and most probably not the same trees, but in some respects, for example carbon storage or the provision of firewood, the impacts for the people will be similar. Or so the story goes, in the countries that have undergone the forest transition. Does this mean forest closure is inevitable?

Is Forest Closure Possible in a Globalized Economy?

Fragments of forest, together with agroforests and agricultural plots, are components of landscape mosaics at the lowest stage of the forest-transition curve. From this point, the progress of agricultural techniques and the abandonment of the less productive and accessible lands lead to 'a progressive adjustment of agriculture to land capability' (Mather and Needle, 1998). However, land sparing thanks to the intensification of agriculture is not generally accompanied by a decline in cropland at the national scale (Rudel *et al.*, 2009). Countries that have benefited from technical progress in agriculture have used it as a comparative advantage to export agricultural products (surplus) rather than to spare land. Only a strong commitment to conservation made by the State and enforced on the ground, with adequate compensations where required, allows a reversal in deforestation trends.

Another factor needs consideration: France's forest transition benefited from a leakage of deforestation made possible thanks to the massive import of wood products from the African colonies (Meyfroidt *et al.*, 2010). The same observation has been made regarding USA and Amazonian forests (Pfaff and Walker, 2010). Thus national forest transitions in western countries might have been made possible only by a leakage of deforestation to the tropics. If this is the case, temperate forests might have regrown at the cost of tropical forests. Most European, North American and Australian countries experienced their forest transition during the mid-1980s (Mather, 1992), after the green revolution of agriculture, and while still protecting their agricultural products from foreign competition.

Developing countries do not benefit from the same conditions. The global demand for timber and wood products for paper and all related productions consistently increases preventing leakages of logging. Energy provision still relies mainly on fuel wood and charcoal, with poor replacement by non-wooden sources of energy, even in urban areas, which increases pressure on forests. The technical advances of the green revolution – chemicals, fertilizers, mechanization, improved seedlings – have poorly spread over family farming in these countries where farmers are confronted with direct competition against high-tech farming from the Northern countries. The same difficulty is raised for industrial development, on which sector-developed countries are put in competition with emergent countries – competition that stopped secondary economic sectors from important development and limited employment. Last but not least, land tenure is often poorly secured, pushing people to over-exploitation rather than long-term sustainable management of natural resources.

The relationship between forest cover and human population is not linear: social change and historical events as well as conservation policies play an important part in this transition (Mather and Needle, 1998; Lambin and Meyfroidt, 2010). For many regions, an evolution of mentalities might also be needed. As long as forests are looked at as dangerous and useless places, people will be willing to convert them to safe and productive ones. Sumatran agroforests in Indonesia are often feared by women because they host tigers and wild boars, and thus they prefer clean monospecific plantations. Forests in the Western Ghats host tigers and elephants and these enter the coffee plantations (Feintrenie *et al.*, 2010; Garcia *et al.*, 2010). Secure access to land and clear cadastre might also slow down forest conversion by allowing farmers to invest in the long term on their land rather than turning to extractive and intensive use of land and forest resources. Globally, the world population has increased faster than cropland (Rudel, 2009), which argues for a forest transition at this global scale. The critical question is then not so much how to slow or halt deforestation and fragmentation, but how we can hasten the transition for developing countries.

References

Agrawal, A. and Ostrom, E. (2001) Collective action, property rights and devolution of forest and protected area management. In: Meinzen-Dick, R.S., Knox, A. and Di Gregorio, M. (eds) *Collective Action, Property Rights and Devolution of Natural Resource Management. Exchange of Knowledge and Implications for Policy*. Food and Agriculture Development Centre (DSE), Feldafing, Germany, pp. 21–25.

Allan, B.F., Keesing, F. and Ostfeld, R.S. (2003) Effect of forest fragmentation on Lyme disease risk. *Conservation Biology* 17, 267–272.

Angelsen, A. and Kaimowitz, D. (eds) (2001) *Agricultural Technologies and Tropical Deforestation*. CAB International, Wallingford, UK <http://www.cifor.org/publications/pdf_files/Books/BAngelsen0101E0.pdf> (accessed 29 July 2014).

Angelsen, A. and Wunder, S. (2003) *Exploring the Forest – Poverty Link: Key Concepts, Issues and Research Implications*. CIFOR Occasional Paper 40. Center for International Forestry Research (CIFOR), Bogor, Indonesia.

Bal, P., Nath, C.D., Nanaya, K.M., Kushalappa, C.G. and Garcia, C. (2011) Elephants also like coffee: trends and drivers of human–elephant conflicts in coffee agroforestry landscapes of Kodagu, Western Ghats, India. *Environmental Management* 47, 789–801.

Barr, C. (2001) *Banking on Sustainability: Structural Adjustment and Forestry Reform in Post-Suharto Indonesia.* WWF Macroeconomics Program Office and Center for International Forestry Research (CIFOR), Washington, DC.

Barua, M., Bhagwat, S.A. and Jadhav, S. (2013) The hidden dimensions of human–wildlife conflict: health impacts, opportunity and transaction costs. *Biological Conservation* 157, 309–316.

Bhagwat, S.A. and Rutte, C. (2006) Sacred groves: potential for biodiversity management. *Frontiers in Ecology and the Environment* 4, 519–524.

Bhagwat, S.A., Kushalappa, C.G., Williams, P.H. and Brown, N. (2005a) A landscape approach to biodiversity conservation of sacred groves in the Western Ghats of India. *Conservation Biology* 19, 1853–1862.

Bhagwat, S.A., Kushalappa, C.G., Williams, P.H. and Brown, N. (2005b) The role of informal protected areas in maintaining biodiversity in the Western Ghats of India. *Ecology and Society* 10, 8.

Brauman, K.A., Daily, G.C., Duarte, T.K. and Mooney, H.A. (2007) The nature and value of ecosystem services: an overview highlighting hydrologic services. *Annual Review of Environment and Resources* 32, 67–98.

Byron, N. and Arnold, M. (1999) What futures for the people of the tropical forests? *World Development* 27, 789–805.

Casson, A. and Obidzinski, K. (2002) From new order to regional autonomy: shifting dynamics of "illegal" logging in Kalimantan, Indonesia. *World Development* 30, 2133–2151.

Chandrakanth, M.G. and Nagaraja, M.G. (1997) Existence value of Kodagu sacred groves: implications for policy. In: Agarwal, A. (ed.) *The Challenge of the Balance: Environmental Economics in India.* Center for Science and Environment, New Delhi, pp. 217–224.

Chapman, C.A., Gillespie, T.R. and Goldberg, T.L. (2005) Primates and the ecology of their infectious diseases: how will anthropogenic change affect host–parasite interactions? *Evolutionary Anthropology: Issues, News and Reviews* 14, 134–144.

Chomitz, K.M., Buys, P., De Luca, G., Thomas, T.S. and Wertz-Kanounnikoff, S. (2007) *At Loggerheads? Agricultural Expansion, Poverty Reduction and Environment in the Tropical Forests.* World Bank, Washington DC.

Cochrane, M.A. and Laurance, W.F. (2002) Fire as a large-scale edge effect in Amazonian forests. *Journal of Tropical Ecology* 18, 311–325.

Colchester, M. (2006) *Justice in the Forest: Rural Livelihoods and Forest Law Enforcement.* Forest Perspectives 3. Center for International Forestry Research (CIFOR) <http://www.cifor.org/publications/pdf_files/Books/BColchester0601.pdf> (accessed 29 July 2014).

Collinge, S.K. (2009) *Ecology of Fragmented Landscapes.* Johns Hopkins University Press, Baltimore, Maryland.

Contreras-Hermosilla, A. (2000) *The Underlying Causes of Forest Decline.* CIFOR Occasional Paper No. 30. Center for International Forestry Research (CIFOR), Bogor, Indonesia.

Crawshaw, P. Jr (2003) A personal view on the depredation of domestic animals by large cats in Brazil. *Natureza Conservaçao* 1, 71–73.

Crome, F. (1997) Researching tropical forest fragmentation: shall we keep on doing what we're doing. In: Laurance, W.F. and Bierregaard, R.O. (eds) *Tropical Forest Remnants: Ecology, Management and Conservation of Fragmented Communities.* University of Chicago Press, Chicago, Illinois, pp. 485–587.

CRP6 (2011) *Forests, Trees and Agroforestry: Livelihoods, Landscapes and Governance.* Center for International Forestry Research (CIFOR), Bogor, Indonesia.

Dale, V.H. and Pearson, S. (1997) Quantifying habitat fragmentation due to land-use change in Amazonia. In: Laurance, W.F. and Bierregaard, R.O. (eds) *Tropical Forest Remnants: Ecology, Management and Conservation of Fragmented Communities.* University of Chicago Press, Chicago, Illinois, pp. 400–409.

Daszak, P., Cunningham, A.A. and Hyatt, A.D. (2001) Anthropogenic environmental change and the emergence of infectious diseases in wildlife. *Acta Tropica* 78, 103–116.

Didham, R.K., Kapos, V. and Ewers, R.M. (2012) Rethinking the conceptual foundations of habitat fragmentation research. *Oikos* 121, 161–170.

Easterly, W. and Levine, R. (1997) Africa's growth tragedy: policies and ethnic divisions. *Quarterly Journal of Economics* 112, 1203–1250.

Ellis, F. (2000) *Rural Livelihoods and Diversity in Developing Countries.* Oxford University Press, Oxford.

Fahrig, L. (2003) Effects of habitat fragmentation on biodiversity. *Annual Review of Ecology, Evolution, and Systematics* 34, 487–515.

Fearnside, P.M. (2005) Deforestation in Brazilian Amazonia: history, rates and consequences. *Conservation Biology* 19, 680–688.

Feintrenie, L., Schwarze, S. and Levang, P. (2010) Are local people conservationists? Analysis of transition dynamics from agroforests to monoculture plantations in Indonesia. *Ecology and Society* 15, 37.
Fischer, J. and Lindenmayer, D.B. (2007) Landscape modification and habitat fragmentation: a synthesis. *Global Ecology and Biogeography* 16, 265–280.
Forman, R.T. (1995) *Land Mosaics: The Ecology of Landscapes and Regions*. Cambridge University Press, Cambridge.
Freeman, J.R. (1999) Gods, groves and the culture of nature in Kerala. *Modern Asian Studies* 33, 257–302.
Fukuyama, F. (2007) Development and the limits of institutional design. In: Dinello, N. and Popov, V. (eds) *Political Institutions and Development: Failed Expectations and Renewed Hopes*. Edward Elgar Publishing, Cheltenham, UK, pp. 21–42.
Gadgil, M. and Vartak, V.D. (1975) Sacred groves of India – a plea for continued conservation. *Bombay Natural History Society Journal* 72, 312–320.
Garcia, C. and Pascal, J.P. (2006) Sacred forests of Kodagu: ecological value and social role. In: Cederlöf, G. and Sivaramakrishnan, K. (eds) *Ecological Nationalisms: Nature, Livelihoods and Identities in South Asia*. University of Washington Press, Seattle, and London, pp. 199–229.
Garcia, C., Marie-Vivien, D., Kushalappa, C.G., Chengappa, P.G. and Nanaya, K.M. (2007) Geographical indications and biodiversity in the Western Ghats, India. Can labeling benefit producers and the environment in a mountain agroforestry landscape? *Mountain Research and Development* 27, 206–210.
Garcia, C.A., Bhagwat, S.A., Ghazoul, J., Nath, C.D., Konerira, M.N., Cheppudira, K.G., Raghuramulu, Y., Nasi, R. and Vaast, P. (2010) Biodiversity conservation in agricultural landscapes: challenges and opportunities of coffee agroforests in the Western Ghats, India. *Conservation Biology* 24, 479–488.
Geist, H.J. and Lambin, E.F. (2002) Proximate causes and underlying driving forces of tropical deforestation. *Bioscience* 52, 143–150.
Geist, H., Lambin, E., Palm, C. and Tomich, T. (2006) Agricultural transitions at dryland and tropical forest margins: actors, scales and trade-offs. In: Brouwer, F. and McCarl, B. (eds) *Agriculture and Climate Beyond 2015*. Springer, The Netherlands, pp. 53–73.
Gillespie, T.R., Chapman, C.A. and Greiner, E.C. (2005) Effects of logging on gastrointestinal parasite infections and infection risk in African primates. *Journal of Applied Ecology* 42, 699–707.
Godoy, R., Reyes-García, V., Byron, E., Leonard, W.R. and Vadez, V. (2005) The effect of market economies on the well-being of indigenous peoples and on their use of renewable natural resources. *Annual Review of Anthropology* 34, 121–138.
Godoy, R., Reyes-García, V., Vadez, V., Leonard, W.R., Tanner, S., Huanca, T. and Wilkie, D. (2009) The relation between forest clearance and household income among native Amazonians: results from the Tsimane' Amazonian panel study, Bolivia. *Ecological Economics* 68, 1864–1871.
Gourlet-Fleury, S., Mortier, F., Fayolle, A., Baya, F., Ouédraogo, D., Bénédet, F. and Picard, N. (2013) Tropical forest recovery from logging: a 24 year silvicultural experiment from Central Africa. *Philosophical Transactions of The Royal Society B: Biological Sciences* 368, 20120302.
Haila, Y. (2002) A conceptual genealogy of fragmentation research: from island biogeography to landscape ecology. *Ecological Applications* 12, 321–334.
Jadhav, S. and Barua, M. (2012) The elephant vanishes: impact of human–elephant conflict on people's wellbeing. *Health and Place* 18, 1356–1365.
Jenks, C. (2004) *Subculture: The Fragmentation of the Social*. Sage Publications, London.
Kahn, M.E. (2000) The environmental impact of suburbanization. *Journal of Policy Analysis and Management* 19, 569–586.
Kaimowitz, D. and Angelsen, A. (1998) *Economic Models of Tropical Deforestation: A Review*. Center for International Forestry Research (CIFOR), Bogor, Indonesia <http://www.cifor.org/publications/pdf_files/Books/Model.pdf> (accessed 29 July 2014).
Kanninen, M., Murdiyarso, D., Seymour, F., Angelsen, A., Wunder, S. and German, L. (2007) *Do Trees Grow On Money? The Implications of Deforestation Research for Policies to Promote REDD*. Forest Perspectives 4. Center for International Forestry Research (CIFOR), Bogor, Indonesia.
Kapos, V. (1989) Effects of isolation on the water status of forest patches in the Brazilian Amazon. *Journal of Tropical Ecology* 5, 173–185.
Kauppi, P.E., Ausubel, J.H., Fang, J., Mather, A.S., Sedjo, R.A. and Waggoner, P.E. (2006) Returning forests analyzed with the forest identity. *Proceedings of the National Academy of Sciences USA* 103, 17574–17579.

Kupfer, J.A., Malanson, G.P. and Franklin, S.B. (2006) Not seeing the ocean for the islands: the mediating influence of matrix-based processes on forest fragmentation effects. *Global Ecology and Biogeography* 15, 8–20.

Kushalappa, C.G. and Bhagwat, S. (2001) Sacred groves: biodiversity, threats and conservation. In: Shaanker, R.U., Ganeshaiah, K.N. and Bawa, K.S. (eds) *Forest Genetic Resources. Status, Threats and Conservation Strategies*. Oxford and IBH Publishing, New Delhi, pp. 21–29.

Lambin, E.F. and Meyfroidt, P. (2010) Land use transitions: socio-ecological feedback versus socio-economic change. *Land Use Policy* 27, 108–118.

Lambin, E.F., Geist, H.J. and Lepers, E. (2003) Dynamics of land-use and land-cover change in tropical regions. *Annual Review of Environment and Resources* 28, 205–241.

Levang, P., Sitorus, S. and Dounias, E. (2007) City life in the midst of the forest: a Punan hunter-gatherer's vision of conservation and development. *Ecology and Society* 12, 18.

Lindenmayer, D.B. and Fischer, J. (2007) Tackling the habitat fragmentation panchreston. *Trends in Ecology and Evolution* 22, 127–132.

MacArthur, R.H. and Wilson, E.O. (1967) *The Theory of Island Biogeography*. Princeton University Press, Princeton, New Jersey.

MacDonald, K. and Rudel, T.K. (2005) Sprawl and forest cover: what is the relationship? *Applied Geography* 25, 67–79.

Malhotra, K.C., Gokhale, Y. and Das, K. (2001) *Sacred Groves of India: An Annotated Bibliography*. Indian National Science Academy and Development Alliance, New Delhi.

Mani, S. and Parthasarathy, N. (2005) Biodiversity assessment of trees in five inland tropical dry evergreen forests of peninsular India. *Systematics and Biodiversity* 3, 1–12.

Martens, P. and Rotmans, J. (2005) Transitions in a globalising world. *Futures* 37, 1133–1144.

Mather, A.S. (1992) The forest transition. *Area* 24, 367–379.

Mather, A.S. (2007) Recent Asian forest transitions in relation to forest-transition theory. *International Forestry Review* 9, 491–502.

Mather, A.S. and Needle, C.L. (1998) The forest transition: a theoretical basis. *Area* 30, 117–124.

McIntyre, S. and Hobbs, R. (1999) A framework for conceptualizing human effects on landscapes and its relevance to management and research models. *Conservation Biology* 13, 1282–1292.

Medjibe, V.P. and Putz, F.E. (2012) Cost comparisons of reduced-impact and conventional logging in the tropics. *Journal of Forest Economics* 18, 242–256.

Megevand, C. (2013) *Deforestation Trends in the Congo Basin*. World Bank, Washington, DC.

Meyfroidt, P., Rudel, T.K. and Lambin, E.F. (2010) Forest transitions, trade and the global displacement of land use. *Proceedings of the National Academy of Sciences USA* 107, 20917–20922.

Michalski, F., Boulhosa, R.L.P., Faria, A. and Peres, C.A. (2006) Human–wildlife conflicts in a fragmented Amazonian forest landscape: determinants of large felid depredation on livestock. *Animal Conservation* 9, 179–188.

Morris, I. (2010) *Why the West Rules – For Now: The Patterns of History and What They Reveal About the Future*. Profile Books, London.

Nasi, R., Brown, D., Wilkie, D., Bennett, E., Tutin, C., Van Tol, G. and Christophersen, T. (2008) *Conservation and Use of Wildlife-based Resources: The Bushmeat Crisis*. CBD Technical Series Paper. Secretariat of the Convention on Biological Diversity, Montreal, Quebec, Canada.

Naughton-Treves, L. (1997) Farming the forest edge: vulnerable places and people around Kibale National Park, Uganda. *Geographical Review* 87, 27–46.

Ormsby, A.A. and Bhagwat, S.A. (2010) Sacred forests of India: a strong tradition of community-based natural resource management. *Environmental Conservation* 37, 320–326.

Ostrom, E. (1990) *Governing the Commons: The Evolution of Institutions for Collective Action*. Cambridge University Press, Cambridge.

Ostrom, E. (1999) *Self-governance and Forest Resources*. Center for International Forestry Research (CIFOR), Bogor, Indonesia.

Pfaff, A. and Walker, R. (2010) Regional interdependence and forest "transitions": substitute deforestation limits the relevance of local reversals. *Land Use Policy* 27, 119–129.

Preston, F.W. (1962) The canonical distribution of commonness and rarity: Part I. *Ecology* 43, 185–215.

Pritchard, B. (2004) Borneo to Broken Hill: how our image of nature affects use. In: Lester, L. and Ellis, C. (eds) *Proceedings of Imaging Nature: Media, Environment and Tourism*. Cradle Mountain, 27–29 June, 2004, University of Tasmania, Hobart, Tasmania, Australia.

Ramakrishnan, P.S., Saxena, K.G. and Chandrashekara, U.M. (eds) (1998) Conserving the sacred: for biodiversity management. Oxford and IBH Publishing Co., New Delhi.

Rudel, T.K. (2009) Tree farms: driving forces and regional patterns in the global expansion of forest plantations. *Land Use Policy* 26, 545–550.

Rudel, T.K., Coomes, O.T., Moran, E., Achard, F., Angelsen, A., Xu, J. and Lambin, E. (2005) Forest transitions: towards a global understanding of land use change. *Global Environmental Change* 15, 23–31.

Rudel, T.K., Schneider, L., Uriarte, M., Turner, B.L. II, DeFries, R., Lawrence, D., Geoghegan, J., Hecht, S., Ickowitz, A. *et al.* (2009) Agricultural intensification and changes in cultivated areas, 1970–2005. *Proceedings of the National Academy of Sciences USA* 106, 20675–20680.

Rutledge, D.T. (2003) *Landscape Indices as Measures of the Effects of Fragmentation: Can Pattern Reflect Process?* DOC Science International Series. Department of Conservation, Wellington, New Zealand.

Sassen, M., Sheil, D., Giller, K.E. and ter Braak, C.J.F. (2013) Complex contexts and dynamic drivers: understanding four decades of forest loss and recovery in an East African protected area. *Biological Conservation* 159, 257–268.

Scanlon, B.R., Jolly, I., Sophocleous, M. and Zhang, L. (2007) Global impacts of conversions from natural to agricultural ecosystems on water resources: quantity versus quality. *Water Resources Research* 43, W03437.

Schure, J., Ingram, V., Marien, J.-N., Nasi, R. and Dubiez, E. (2011) Woodfuel for urban centres in the Democratic Republic of Congo. Center for International Forestry Research (CIFOR), Bogor, Indonesia <http://www.cifor.org/publications/pdf_files/infobrief/3678-brief.pdf> (accessed 29 July 2014).

Shackleton, C.M., Shackleton, S.E., Buiten, E. and Bird, N. (2007) The importance of dry woodlands and forests in rural livelihoods and poverty alleviation in South Africa. *Forest Policy and Economics* 9, 558–577.

Sunderlin, W.D., Angelsen, A., Belcher, B., Burgers, P., Nasi, R., Santoso, L. and Wunder, S. (2005) Livelihoods, forests and conservation in developing countries: an overview. *World Development* 33, 1383–1402.

Tejaswi, G. and Holmgren, P. (2007) *Manual on Deforestation, Degradation and Fragmentation Using Remote Sensing and GIS*. MAR-SFM Working Paper. Forestry Department, Food and Agriculture Organization of the United Nations, Rome.

Thompson, I.D., Guariguata, M.R., Okabe, K., Bahamondez, C., Nasi, R., Heymell, V. and Sabogal, C. (2013) An operational framework for defining and monitoring forest degradation. *Ecology and Society* 18, 20.

Tole, L. (2006) Measurement and management of human-induced patterns of forest fragmentation: a case study. *Environmental Management* 37, 788–801.

Turner, I.M. and Corlett, R.T. (1996) The conservation value of small, isolated fragments of lowland tropical rain forest. *Tree* 11, 330–333.

Turner, M.G. (1989) Landscape ecology: the effect of pattern on process. *Annual Review of Ecology and Systematics* 20, 171–197.

Urech, Z.L., Felber, H.R. and Sorg, J.P. (2012) Who wants to conserve remaining forest fragments in the Manompana corridor? *Madagascar Conservation and Development* 7, 135–143.

Wickham, J.D., Riitters, K.H., Wade, T.G. and Coulston, J.W. (2007) Temporal change in forest fragmentation at multiple scales. *Landscape Ecology* 22, 481–489.

Wilcove, D.S., McLellan, C.H. and Dobson, A.P. (1986) Habitat fragmentation in the temperate zone. In: Soulé, M.E. (ed.) *Conservation Biology: The Science of Scarcity and Diversity*. Sinauer Associates, Sunderland, Massachusetts, pp. 237–256.

Williams, M. (2003) *Deforesting the Earth: From Prehistory to Global Crisis*. University of Chicago Press, Chicago, Illinois.

Wunder, S. and Verbist, B. (2003) *The Impact of Trade and Macroeconomic Policies on Frontier Deforestation*. World Agroforestry Center–ICRAF, Bogor, Indonesia.

10 Governance Failures and the Fragmentation of Tropical Forests

Christopher Barr,[1]* Keith Barney[2] and Sarah A. Laird[3]
[1] *Woods & Wayside International, USA;* [2] *Australian National University, Canberra, Australia;* [3] *People and Plants International, USA*

Abstract

Across the tropics, high rates of forest degradation and deforestation have resulted not only in the loss of large areas of natural forest cover but also in increasing fragmentation of the forests that remain. This chapter examines the underlying governance drivers of forest fragmentation through five key analytical approaches: (i) understanding of historically contested property rights over forests involving states, private-sector actors and communities; (ii) the widespread failure of selective logging timber concessions to achieve sustainability objectives in the tropics; (iii) related problems with weak law enforcement, corruption and patronage politics, and the dominance of commercial forestry by political and military elites; (iv) the conversion of forested areas to other land uses, often in response to market and governance failures; and (v) the globalization of resource commodity networks. Significantly, in considering how extra-sectoral land-use conversion influences forest management, we have found that forest fragmentation is often an intentional, yet typically understated, objective of government spatial planning processes. In some cases, forest fragmentation is even implicit in the approach of certain conservation mechanisms such as the designation of 'high-conservation-value forests'. Forest fragmentation can also be understood not only in biophysical terms but as a socio-cultural phenomenon with profound implications for displaced indigenous groups and other communities living in and around forests. Continued forest fragmentation poses significant challenges for conservation, for the socio-cultural viability of forest-dependent communities and for market-based global forest governance initiatives such as REDD+ (reducing emissions from deforestation and forest degradation).

Introduction

Across the tropics, high rates of forest degradation and deforestation have resulted not only in the loss of large areas of natural forest cover but also in increasing fragmentation of the forests that remain. Forest fragmentation – broadly defined as 'any process that results in the conversion of formerly continuous forest into patches of forest separated by non-forested lands' (Convention on Biological Diversity, 2014) – can occur at multiple scales, ranging from a single forest stand to a landscape defined by biophysical or administrative boundaries, or even a bioregion stretching across national borders. Fragmentation has long been recognized as a potent threat to the continued functional integrity of tropical forest

*cbarr@woods-wayside.org

ecosystems and the biodiversity they sustain (Laurance *et al.*, 1998; Bierregaard *et al.*, 2001). For at least three decades, forest ecologists and conservationists have warned that fragmented forests often have limited capacity to support critical species of flora and fauna, and this diminishes not only the biological richness of these forests but also the environmental services they provide and the livelihoods they support (Harris, 1984). In recent years, forest fragmentation has also posed significant technical and institutional challenges for the United Nations (UN)-led initiative of REDD+ (reducing emissions from deforestation and forest degradation) and related efforts to mitigate climate change.

In much of the tropics, processes of forest fragmentation have been catalysed initially by commercial logging and have accelerated as logged-over forests have been converted for smallholder agriculture, commercial tree plantations and agro-industrial crops. Large areas designated as state production forests have been subject to multiple rounds of logging, leaving them depleted of commercial species, biologically degraded with limited capacity for regeneration and vulnerable to a host of continuing threats. In particular, logging roads running through former concession sites provide ready access to informal loggers seeking residual trees, to settlers expanding the agricultural frontier and to corporate actors developing new areas of cropland and pasture. In recent years, growing demand in global markets for commodities including pulp, oil palm, soybean, beef, sugarcane and para-rubber – as well as a scarcity of arable land in some emerging consumer countries – has created powerful incentives for the conversion of natural forests to industrial plantations and different non-forestry land uses. As tropical forest landscapes have become increasingly fragmented, stands of mature primary forests often remain intact only where extraction of timber and/or conversion to alternate land uses are not yet commercially feasible.

As with the broader processes of deforestation and degradation, forest fragmentation is often driven by weaknesses, contradictions and outright failures of governance institutions. We define 'forest governance' as the institutional processes and structures, and modes of political power and authority, that shape 'how decisions are made about the management and use of forest lands and resources' (Box 10.1). We note that decisions about forests are significantly influenced by governance institutions and processes situated both within and outside the forestry sector. In this chapter, we will examine the links between forest fragmentation and governance failures, and examine why institutions and processes established to administer forests – including those charged with ensuring sustainable forest management and forest protection – are so frequently undermined by a variety of factors that produce and reinforce weak governance.

Box 10.1. What is forest governance?

The literature on forest governance offers a variety of definitions. For the purposes of this chapter, we use the following definition, formulated by the World Resources Institute:

> Forest governance refers to how decisions are made about the management and use of forest lands and resources. These decisions are shaped and influenced by a complicated web of actors, rules, and practices both within and beyond the forest sector. Poor forest governance is often characterized by low levels of transparency, accountability, and public participation in decision making, as well as weak coordination across different sectors and levels of government. It contributes to problems such as unplanned forest conversion, conflict over forest resources, illegal logging, and corruption. These underlying governance challenges remain a persistent obstacle to achieving sustainable and equitable management of forests in many countries.

(From World Resources Institute Governance of Forests Initiative 2013, http://www.wri.org/project/governance-of-forests-initiative.)

As we will explore in this chapter, timber concession regulations often allow logging to occur at levels of intensity that lead to rapid commercial and biological depletion, rather than long-term sustainable management across multiple harvesting rotations. Moreover, timber companies have strong economic incentives to exceed harvesting limits and to avoid making investments in sustainable management of their concession areas. In many countries, state leaders have distributed timber concessions to companies associated with powerful government and military officials as a form of patronage, and corruption within agencies administering forest resources is widespread. In such contexts, state forestry agencies often have limited capacity or political will to enforce their own forestry laws, effectively rendering large areas of forest land as an 'open access' resource with few constraints on extra-legal timber harvesting.

We will examine how fragmentation has been linked to the conversion of forestlands to other land uses, and in this way, has been driven by policies, institutional structures and market signals from outside the forestry sector. Government policy and planning decisions aimed at enhancing the profitability of land-extensive commodities often drive forest fragmentation by providing incentives for agricultural expansion in frontier areas. Macroeconomic policies can similarly encourage deforestation when they improve the terms of trade for commodities such as soy, beef or oil palm. As global and regional markets have become increasingly integrated in recent decades, deforestation and fragmentation in many tropical countries have been driven by growing demand for key commodities in China and other major consumer countries. Globalization has, in turn, connected tropical forest landscapes – and the governance of these resources – to new commodity chains, new sources of capital investment and new transnational corporate actors.

Significantly, governance failures driving the fragmentation of tropical forests have also led to the displacement of customary management systems practised by forest-reliant communities. In this way, forest fragmentation can be understood not only in biophysical terms but also as a socio-cultural phenomenon with profound implications for indigenous groups and other communities living in and around forests. The fragmentation of tropical forests underscores a pressing need for innovative and effective governance mechanisms. We will argue that these are required, on the one hand, to slow or reverse the pace of both biophysical and socio-cultural fragmentation, and on the other hand, to address the specific challenges associated with managing fragmented forest resources.

The 'Political Forest' and Commercial Timber Concessions

Approximately 75% of the world's forests are controlled by governments (Sunderlin *et al.*, 2008). In many tropical countries – including Indonesia, Democratic Republic of Congo, Angola, Venezuela, Zambia, Myanmar, Laos, Cambodia, Central African Republic, Congo, Gabon, Cameroon and Mozambique – virtually all forested lands are administered by state agencies (Sunderlin *et al.*, 2008). In some tropical countries – including Brazil, Peru, Colombia and Bolivia – management control of significant areas has been devolved to indigenous peoples, rural communities and/or private companies (Sunderlin *et al.*, 2008). Across the tropics, however, government forestry departments are charged with maintaining a 'permanent forest estate', portions of which are designated for commercial exploitation, conservation, watershed protection and other environmental services, management by rural communities, or other authorized and zoned uses. In many tropical countries, state forestry agencies and the legal regulatory structures they use to administer forests were initially established during the colonial era, although these have often been adapted and vastly expanded by post-colonial states.

Tracing the origins of what they term the 'political forest', Peluso and Vandergeest (2001) argued that, in South-east Asia, the idea of state territorial sovereignty over a category of land identified as legal 'forests' emerged in the 19th century. During this period, the region's colonial governments asserted control over forested landscapes and resources, and placed these under the administrative authority of state forestry bureaucracies. The creation of the political forest commonly included: 'territorial zoning and mapping, the enactment of forest laws delimiting legal and illegal forest uses, the constitution of state forestry institutions to implement these laws according to specific procedures, the constitution of forest police, and the creation of legal exemptions that became Customary Rights' (Peluso and Vandergeest, 2001, pp. 764–765). Importantly, the assertion of state control over forests meant that 'Access to forested lands or particular species was made contingent on government-issued permits' (Peluso and Vandergeest, 2001, p. 765). Although the specific institutions and practices used to administer forests varied across political systems and geographical contexts, successive colonial and post-colonial governments carried out similar projects in much of the tropics. These political forests served as a legal mechanism for the delegitimation and exclusion of previously existing forest-land management systems and livelihoods.

In further institutionalizing the political forest, many post-colonial forestry departments – often in partnership with key global organizations such as the UN Food and Agriculture Organization (FAO) – have classified large areas of territory as 'production forests' and made these exclusively available to commercial logging by public and private timber companies. In overseeing the commercial exploitation of forests, government forestry agencies have frequently pursued multiple economic and political objectives. On the one hand, they have sought to promote national economic development by encouraging investment in the sector and by facilitating the production of timber and other commodities for domestic and international markets. On the other hand, they have generated revenue flows for the government itself through licensing fees, timber royalties, export taxes and other forest fiscal payments.

Although timber concession regulations vary among tropical forest countries, they are generally structured to allow licence-holders to harvest timber selectively from a specified area under approved management plans for a defined period of time. Rooted in the principles of scientific forestry, selective logging concessions were initially designed by colonial forestry services to ensure sustained yields of high-value species, such as teak or mahogany. For several decades, concessionary logging has been promoted by the FAO, the World Bank and other international organizations as the primary institutional structure for achieving sustainable forest management (Vandergeest and Peluso, 2006; Independent Evaluation Group, 2012).

Globally, this conflation of managed selective logging and sustainable forest management has been widely embraced by a diverse range of institutional actors (Zimmerman and Kormos, 2012). Indeed, it is by far the dominant view, shared by professional forestry organizations, bilateral and multilateral donor institutions, forestry research institutes and major conservation organizations.[1] Most recently, the UN REDD+ initiative has promoted the creation of incentives for sustainable forest management through managed selective logging as part of its broader efforts to reduce deforestation and degradation of tropical forests (Zimmerman and Kormos, 2012). As elaborated below, however, this broad acceptance of selective logging timber concessions as the primary institutional mechanism for sustainable forest management is increasingly at odds with the documented long-term performance of such systems in much of the tropics.

Timber Rents and the Breakdown of Sustainable Logging

Paradoxically, the widespread failure of commercial timber concessions to achieve sustainability objectives has been a major catalyst of forest fragmentation processes in much of the tropics. In theory, most timber concession systems are modelled around selective harvesting, and are structured to allow logged-over areas to regenerate before the next rotational felling occurs to facilitate sustained productivity of forests over the long term. In this way, commercial forestry regulations commonly require timber companies to manage their concessions according to the following principles:

- Rotational harvesting of concession areas.
- Minimum felling diameters.
- Area-based restrictions on harvesting intensity.
- Restrictions on extraction of protected species.
- Retention of minimal seed trees.
- Enrichment replanting of logging sites.

Numerous initiatives have sought to enhance the performance of selective logging concessions systems by introducing policies that promote specialized harvesting practices for reduced-impact logging, more efficient technologies to avoid waste, the use of performance bonds and area fees to guarantee adherence to management plans and long-term tenure security for large-scale forestry concessions (Repetto and Gillis, 1988; World Bank, 1992; Sist *et al.*, 1998; Kaimowitz, 2004; Boscolo and Vincent, 2007).

Over the past three decades, however, a growing literature from Asia, Africa and Latin America has documented a widespread and continuous failure of commercial timber concessions to achieve stated sustainability goals. In a recent review, Zimmerman and Kormos (2012, p. 479) argued that 'a convincing body of evidence shows that as it is presently codified, sustainable forest management (SFM) logging implemented at an industrial scale guarantees commercial and biological depletion of high-value timber species within three harvests in all three major tropical forest regions.' Specifically, they concluded that concession regulations in most timber-producing countries allow logging to occur at levels of intensity that result in permanent damage to the forest canopy and to residual trees, and that suppress regeneration of primary forest tree species:

> Under prevailing tropical SFM protocols, (a) minimum rotation cycles, typically 25–35 years, are too short by a factor of at least two; (b) minimum felling diameters, most often 50 cm DBH [diameter at breast height], are too small to preserve adequate populations of reproductive adults and to control harvest intensity, assuming that diameter-limit cutting is the only option; (c) per-unit-area harvest intensities (around eight trees per ha) are too high and leave residual stands most often lacking future tree crop regeneration, overgrown by pioneer vegetation, and fire prone, especially in regions that experience dry seasons; and (d) seed-tree retention rates – that for Brazil is 10% retention of trees greater than 45 cm DBH – are inadequate.
>
> (Zimmerman and Kormos, 2012, p. 483)

Even where reduced-impact logging techniques have been used to decrease collateral damage within residual stands, Zimmerman and Kormos (2012, p. 482) found that 'high logging intensities and low minimum cutting diameters usually left such a high percentage of the canopy in open gap (20%–50%) that the residual stand became colonized by fast-growing, light-loving vines and pioneer species…suppress[ing] the regeneration of slower-growing, shade-tolerant and shade-intolerant primary forest trees.'

Although challenged by some analysts, the conclusions reached by Zimmerman and Kormos (2012) are both striking and disconcerting: they suggest that, in most contexts, commercial logging companies are unlikely to achieve either biological or commercial sustainability, even if they adhere fully to timber concession regulations. To complicate matters, the institutional breakdown of sustainable forest management through industrial logging is hardly limited to poorly formulated rotation cycles, inadequate harvesting limits, and other technical weaknesses in concession guidelines. Since the mid-1980s, rather, forest economists have shown that logging companies also have strong economic incentives to disregard forest management plans and to refrain from making any significant investments in the long-term sustainability of the forestry concessions they manage.

Repetto and Gillis (1988) documented how governments in Indonesia, Malaysia, the Philippines, Ghana and Brazil have maintained timber royalties, concession fees and other components of forest fiscal systems at levels that allow logging companies to capture economic rents – that is, profits above a normal rate of return. They and others argued that, by undervaluing forest resources in this way, governments not only fail to collect significant amounts of potential revenues, but, in doing so, they also effectively encourage timber companies to manage their concessions in ways that maximize profits over the short term (Ascher, 1999; Ross, 2001; Barbier, 2005). Concession-holders often recognize that their access to economic rents from state-administered forests – ostensibly a publicly owned resource – may last for only a limited time due to the expiration of concession contracts, potential policy changes and/or physical depletion of high-value timber species. Faced with such uncertainties, they have strong incentives to extract as much value as possible from the forests they manage while it is available. As Barbier *et al.* (2010, p. 103) noted, 'Such resource-extractive activities tend to be short-term, high-payoff investments that are highly destructive to forests and ecosystems.'

Analysing these trade-offs in lowland mahogany forests of Bolivia, Rice *et al.* (1997, p. 46) discovered that 'unrestricted logging is from two to five times more profitable than logging in a way that would ensure a continued supply of mahogany.' They argued that similar dynamics drive unsustainable logging across the tropics, as timber companies often have 'no economic incentive to invest in sustainable management':

> Logging, as typically practiced in the tropics, rapidly harvests the most highly valued trees. The number of species may be as low as one (where there is a specialty wood, such as mahogany) or as high as 80 or 90 (where there is demand for high variety). Logging companies generally show little concern for the condition of residual stands and make no investment in regeneration. This attitude emerges, in part, as a matter of simple economics. In deciding whether to restrict harvests, companies face a choice between cutting trees immediately and banking the profits or delaying the harvest and allowing the stand to grow in volume and value over time. Economics, it seems, dictates the decision.
>
> (Rice *et al.*, 1997, p. 46)

Interest rates are a significant factor shaping this decision. As Rice *et al.* (1997, p. 46) explained, 'Because risks are considerable and capital is scarce, real interest rates in developing countries are often much higher than in industrial countries[2].... Thus, companies that rapidly harvest their assets can invest their profits immediately and generate continuing rates of return. The benefits of delaying harvest, in contrast, are small.' Indeed, the slow growth rates of most high-value species, combined with uncertainties concerning future prices and mortality risks, give logging companies little incentive to manage their concessions for optimal profits 40 or 80 years in the future (Karsenty, 2000; Kaimowitz, 2004). Rice and colleagues noted that 'like most other business people, who are unwilling to make risky investments in developing countries unless offered considerably higher returns, loggers choose to cut their trees as quickly as they can' (Rice *et al.*, 1997, p. 46).

As high-value species are depleted and formal access to timber rents declines, concession-holders often have an even greater incentive to log outside of approved harvesting sites, to cut trees below the legal minimum diameter and to engage in any number of other violations. Describing the declining profitability of forestry concessions in Indonesia during the late 1990s, Barr (2001a) noted that logging companies have little motivation to play by the rules if doing so provides only marginal or negative returns on their investment. On the contrary, timber companies have a strong incentive to overharvest their concession sites when government forestry departments demonstrate a willingness, as they do in Indonesia, to reclassify areas of production forests for full-scale clearing once they reach a designated threshold of degradation (Meijard and Sheil, 2007; Barr and Sayer, 2012).

In this way, governance failures related to commercial timber extraction frequently represent a critical first step in a multistage process leading to fragmentation of tropical forests on a large scale. Although framed by the discourse of sustainability, empirical studies indicate that industrial logging has led to the degradation of forest ecosystems through overharvesting of concession sites and a failure to support regeneration of residual stands (Karsenty and Gourlet-Fleury, 2006; Cerutti *et al.*, 2008; Clark *et al.*, 2009;

Nasi and Frost, 2009). Many logged-over concession sites also have significant volumes of dead biomass, making them susceptible to fire, and the loss of canopy makes them vulnerable to invasive species, such as *Imperata* and other grasses. With logging roads and other infrastructure developments providing access to previously remote areas, residual forests within logged-over concessions are also susceptible to conversion by smallholder settlers and commercial land users expanding the agricultural frontier.

Weak Law Enforcement, Corruption and Patronage Politics

Weak forest law enforcement is a major factor facilitating uncontrolled logging both within and outside areas designated for commercial timber production. Numerous studies have sought to quantify the scale of illegal logging and illegal trade of wood products at both national and global levels; however, estimates vary widely and are, unavoidably, based on partial evidence and assumptions (Box 10.2). A 2004 study concluded that the global value of illegally produced roundwood, lumber and plywood amounted to US$22.5 billion on an annual basis (Seneca Creek Associates and Wood Resources International, 2004). Of this, approximately US$5.0 billion is traded internationally, amounting to 7–10% of world exports of primary wood products. A more recent analysis estimates that illegal harvesting accounts for 70% of logging in the Brazilian Amazon, 60% in Ghana and Indonesia, 35% in Cameroon and 25% in Malaysia (Lawson and MacFaul, 2010). The World Bank has estimated that governments lose US$10–15 billion annually due to forgone revenues associated with illegal logging and trade (World Bank, 2002; Goncalves *et al.*, 2012).

In many tropical countries, the scale of illegal logging and associated trade underscores a fundamental paradox: state forestry agencies have established far-reaching authority to administer forest territories and the resources they contain, yet their performance is frequently constrained by limited budgets, weak law enforcement capacity, competing bureaucratic agendas and, in many cases, powerful political disincentives for sustainable management (Karsenty and Ongolo, 2011). Indeed, many tropical forest countries can be characterized as 'fragile states' – defined by the Organisation for Economic Co-operation and Development (OECD) as those countries where there is a 'lack [of] political will and/or capacity to provide the basic functions needed for poverty reduction, development and to safeguard the security and human rights of their populations' (OECD/DAC, 2007, p. 2; Irland, 2008).

Box 10.2. Illegal logging

Illegal logging is commonly defined as occurring 'when timber is harvested, transported, bought or sold in violation of relevant national laws' (Lawson and MacFaul, 2010). Among the most significant practices associated with illegal timber harvesting and trade are the following:

- Logging in protected forests or other areas not designated for timber harvesting.
- Logging in production forests by unauthorized actors.
- Logging by licensed concession-holders in ways that violate approved management plans, including: (i) harvesting outside of designated felling sites; (ii) harvesting volumes of timber in excess of approved levels; and (iii) harvesting trees below minimum diameter limits.
- Logging or trade of protected species.
- Under-reporting of volumes harvested, transported, or traded to avoid royalty or tax payments.
- Transporting, exporting, or importing timber without required permits.

The specific practices that constitute illegal timber harvesting and trade vary widely across jurisdictions: what is legal in one country can be illegal in another.

(From Lawson and MacFaul, 2010; Seneca Creek Associates and Wood Resources International, 2004.)

Examining the effects of state fragility on forest governance, Harwell (2010, p. 5) further elaborated:

> Although not explicit, the definition implies that fragile states lack rule of law (not just monopoly over the use of force), strong institutions, control of private interests and independence of the state from them, and accountability and concern for [the] well-being of [their] population. These weaknesses lead to a context of illegitimate and dysfunctional government and vulnerable citizenry that is not conducive to forest conservation and management.

Corruption, a common characteristic of fragile states, is widespread among tropical forest countries, significantly undermining the ability of governments to enforce their own forestry laws (Irland, 2008; Harwell, 2009; Cerutti *et al.*, 2013). At the highest levels, grand corruption can be a significant driver of forest fragmentation when it involves, for instance, bribery of senior officials by corporate actors seeking to obtain timber concessions or forest conversion licences, influence peddling over land-use policy or planning decisions, or avoidance of prosecution for illegal clearing of forests. At the field level, petty corruption can have similar effects (albeit on a smaller scale) when timber companies make informal payments to local forestry officials to approve harvesting permits, ignore violations at logging sites or allow trucks carrying illegal timber to pass government checkpoints. Where illegally harvested logs are destined for export markets, corruption among customs officials, and a lack of due diligence and interest in consumer countries, frequently facilitates the flow of illicit timber and wood products through global commodity chains.

In analysing the effects of state fragility on forest governance, Harwell (2010) distinguished between fragile states that are relatively strong and those that are relatively weak. In the 'stronger' fragile states (also referred to as 'predatory states'), political leaders commonly use timber concessions as a form of patronage, distributing valuable logging licences to senior government officials and military officers to secure their loyalty and support. In these circumstances, Harwell (2010, p. 6) argues:

> Such states are not 'incapable'; rather, the elite choose to focus their capacities not on the pursuit of the public good but on consolidating personal and state power through patronage relationships with the private sector and a wide array of specialized criminal networks. This patronage undermines the rule of law, encourages the expansion of shadow economies, and… further strengthen[s] criminal networks.

Indonesia's Soeharto regime (*c.*1966–1998), for instance, allocated licences to millions of hectares of timberland to companies affiliated with senior generals, regional military commanders, state ministers, governors and members of the president's own family (Barr, 1998; Brown, 1999). In doing so, the regime's leaders established a high level of personal allegiance, linked to an ongoing flow of economic benefits, among powerful actors at all levels of the state apparatus (Ross, 2001). During the initial decade of the regime, these actors were allowed to extract timber rents from their concessions with relatively minimal oversight of their operations by the Ministry of Forestry. By the mid-1980s, however, as Soeharto had consolidated his power, the state banned the export of raw logs and required concession holders to make downstream investments in plywood production. Access to export markets – and by extension, forest rents – was then tightly controlled by the Indonesian Wood Panel Producers Association (APKINDO), a powerful cartel managed by an associate of Soeharto (Barr, 1998).

Similarly, in Liberia, the government of Charles Taylor used timber concessions as patronage, both to enrich the regime's senior leaders and to finance pro-government forces in that country's civil war. Shortly after coming to power, the Taylor regime 're-align[ed] forest lands into several large concessions [and] That "mega-concession" policy allowed the inner circle of the country's leadership to hand out concession rights to favored political cronies, militia leaders, and arms dealers' (Blundell, 2010, p. 14). Several recipients of concession licences allegedly traded timber for weapons and facilitated the import and distribution of arms to combatants in Liberia's civil war, as well as the conflict in neighbouring Sierra Leone. A 2005 review of

Liberia's forestry concessions found that the volumes of timber harvested were grossly under-reported, allowing companies to evade forestry taxes; and large volumes of logs were smuggled to export markets, especially through Côte d'Ivoire. Significantly, the review also found that by the end of the civil war in 2003, the total area of forestry concessions allocated was 2.5 times greater than the nation's actual forest area (Blundell, 2010).

In stronger fragile states, the pervasive presence of government and military elites in timber production has often meant that forestry bureaucracies have little, if any, political will to enforce regulations concerning sustainable management. On the contrary, the strategic significance of timber rents as political patronage has frequently meant that forestry officials are charged with administering forests in ways that advance the accumulation strategies of state elites and their business associates. Analysing the political economy of timber production in Indonesia, Malaysia and the Philippines, Ross (2001) argued that control over the distribution of forest rents – a process he termed 'rent seizing' – is often the subject of intense political struggles among factions within the state.[3] Not surprisingly, where the distribution of timber rents plays a critical role in supporting a regime's political power structure, factions promoting the conservation and sustainable management of forests are often marginalized. This, in turn, shapes what policies are set, who has access to forests, how forest regulations are enforced and, ultimately, how forests are (mis)managed.

By contrast, fragile states are described as being weak when there are significant 'sub-national areas with no state presence and forests are uncontrolled by government authority' (Harwell, 2010, p. 6). The degree of state fragility can vary widely within a particular country's borders and, in regions where fragile states are weakest, there is frequently a negative correlation between levels of state control and density of forest cover. In many forested regions that lie beyond the full reach of state control, forests are managed by local communities under customary tenure regimes. In some such regions, particularly where forests are rich in high-value timber species, forest resources can be controlled and heavily exploited by private militias, criminal syndicates, armed insurgencies and other armed groups. Logging under these circumstances is rarely constrained by concerns over sustainability and, in many cases, it is linked to violent conflict (Harwell, 2010).

Distinctions between strong and weak fragile states are often complicated further in practice. For example, in Laos (Lao People's Democratic Republic), often described as a relatively weak state, the central Communist Party apparatus nevertheless wields significant political authority and control. Through the Lao state's strategy of 'turning land into capital', large-scale land concession deals have been awarded to private developers in exchange for commitments to finance urban infrastructure projects like shopping malls and an international athlete's village (Kenney-Lazar, 2010; Dwyer, 2013). Provincial governments in Laos have also long used this strategy to finance the construction of roads and government office buildings (Hodgdon, 2008). These barter deals have been completed under less-than transparent arrangements, and some have argued that there is also significant rent seeking associated with them (Global Witness, 2013). Timber harvesting and export quotas have also reportedly been allocated to external logging interests in order to repay loans reportedly extended to Laos by the Soviet Union and Vietnam during the Second Indochina War. In these ways, what appears at first glance to be a rather chaotic forestry policy in Laos becomes a more logical strategy adopted by the Lao state to utilize its natural resource endowments to advance elite-defined development objectives (Dwyer, 2013). Such politically defined land deals and timber quota allocations have major implications for patterns of forest fragmentation in Laos.

Illegal logging and associated trade are widely described as being significant drivers of unsustainable forestry in both tropical and temperate regions. However, the links between illegality, forest degradation and forest fragmentation are not always straightforward. While unlicensed loggers (such as groups of villagers with chainsaws) may operate without government harvesting permits and approved management plans, there are examples where their practices may be relatively 'sustainable' as based on selective and low-impact logging practices (McElwee, 2004). Nevertheless, high rates of illegal logging often suggest that both the pace of forest degradation and the areas of forest affected will exceed those anticipated by forestry department planners.

Where illegal logging occurs within the boundaries of existing timber concessions, it can mean that commercial and biological depletion – and by extension, degradation and fragmentation – of production forests occur much faster than they would if these sites were exploited according to concession management plans. Where illegal logging occurs outside of concession boundaries, it can mean that degradation and fragmentation occur on a much broader scale across the national forest estate, well beyond areas designated for commercial production. In contexts where the volumes of illegally harvested timber are sufficient to suppress market prices, widespread illegal logging can also serve as a strong disincentive for more responsible forestry companies to invest in sustainable management, and this can operate at an international scale (Seneca Creek Associates and Wood Resources International, 2004).

The large volumes of illegally harvested timber flowing from the tropics suggest that efforts to reduce the fragmentation of tropical forests will need to include effective mechanisms to curb illegal logging and trade. Since the mid-2000s, numerous initiatives have focused on developing systems for verifying the legality of timber and tracking logs through a chain of custody from the stump to destination markets. For instance, through the Forest Law Enforcement, Governance and Trade (FLEGT) programme, the European Union (EU) has negotiated voluntary partnership agreements (VPAs) with several timber-producing countries, through which wood-based imports are limited to legal products verified through an audit-based licensing scheme (Brack, 2012).[4] Other initiatives aimed at curbing illegal logging and trade have focused on 'following the money' or 'following the investors' (Setiono and Husein, 2005; Global Witness, 2009). Such efforts have sought to apply anti-money laundering laws and asset confiscation regimes to curtail the flow of profits from illegal forestry activities, especially in cases involving powerful actors with bank accounts and other assets in offshore jurisdictions (Goncalves *et al.*, 2012).

Governance Drivers of Land-use Conversion, Fragmentation and Deforestation

The discussion thus far has focused on the links between fragmentation and the degradation of forests resulting from licensed and unlicensed timber extraction. We will now examine the relationship between forest fragmentation and deforestation resulting from the conversion of forested areas to other land uses. On a broad scale, such conversion processes are a significant component of forest fragmentation in the tropics, as they make the boundaries between fragments of mature forests and non-forestry land uses both more expansive and more immutable. As with degradation caused by unsustainable logging practices, tropical deforestation – and the resulting fragmentation of forest landscapes – is frequently catalysed by weak regulations, contradictory policy and planning decisions, market distortions and other governance failures.

In recent decades, agricultural expansion has been the largest driver of deforestation in the tropics. During 1980–2000, more than three-quarters of new crop and pastoral land in these regions was developed on areas covered by forests – some 50% involved the conversion of intact forests and another 28% the clearance of disturbed forests (Gibbs *et al.*, 2010). For the period 2000–2010, agricultural expansion was directly responsible for 73% of tropical deforestation, with commercial agriculture accounting for 40% of total forest loss and local/subsistence agriculture accounting for 33% (Hosonuma *et al.*, 2012).

The relative significance of commercial agribusiness actors versus smallholder agriculturalists as drivers of forest change shows considerable geographical variation. Since the early 1990s, agribusinesses producing beef and soybeans for international markets have been leading drivers of deforestation in the Amazon, and export-oriented production of oil palm (for food, household and cosmetic products) and pulp/paper have been major drivers in South-east Asia (Rudel *et al.*, 2009; Boucher *et al.*, 2011; Obidzinski and Dermawan, 2012).[5] Increasingly, growing demand for biofuels, spurred on by government policies in the USA, Europe and elsewhere, is shaping forest conversion in both regions (e.g. sugarcane in Brazil and Cambodia; oil palm in Malaysia, Indonesia and Colombia). By contrast, expansion of small-scale subsistence agriculture continues to be the leading driver of forest cover loss in much of Africa (Fisher, 2010). As Hosonuma *et al.* (2012, p. 8) noted, however, corporate actors are rapidly gaining significance as agents of forest change in that region: 'While the four African countries with the largest forest areas (Democratic Republic of Congo, Angola, Zambia, and Mozambique) (FAO 2011) are still in the pre- and early-transition phase, forest loss rates and the influence of commercial globalized agriculture are expected to increase...'

In countries where states exert strong control over forest resources, processes of forest conversion – and the resulting patterns of fragmentation – are shaped heavily by government policy and planning decisions. Indeed, we note that forest fragmentation is often an *intentional*, yet unstated, objective of government spatial planning processes. In some cases, states use fragmentation as a means to exercise internal control over state territory and/or to delineate and regulate national boundaries (Vandergeest and Peluso, 1995; Scott, 1998). State-led spatial planning policies effectively determine which portions of the forest estate will remain under forest cover and which will be converted to other land uses.

In sectoral planning processes, state forestry agencies in many countries have demonstrated limited capacity or political will to rehabilitate, manage or protect forests once they are depleted of high-value timber (Barr and Sayer, 2012). Rather, as they have become administrative stewards of increasingly large areas of logged-over forests, forestry departments have often reclassified 'degraded' forestlands (or variations such as 'barren' and 'wasteland') for conversion to monoculture timber or pulpwood plantations, agribusiness estates, infrastructure projects and/or smallholder agricultural settlements (Barr and Sayer, 2012). Here, Barney (2008, p. 101) noted, 'Identifying forest-land as degraded forest is useful for the political project of evacuating these zones of substantive social, economic, and ecological significance.'

In multisectoral spatial planning processes, 'degraded' forests can also be the subject of intense inter-agency struggles, as ministries of forestry, agriculture, mining, industry and infrastructure compete for authority over these areas to advance their own sector's development plans and to control the economic rents associated with the land, residual timber and/or subsoil minerals (Ascher, 1999). Powerful industry lobbies frequently use their political influence (both formal and informal) to shape the policy and planning decisions that emerge from such processes (Ascher, 1999; Barbier, 2005).

In countries where states exercise relatively weak control over forests, pressures for forest conversion are shaped much more heavily by the land-use decisions of smallholders and corporate actors operating in loosely regulated frontier regions (Tsing, 2004; Barney, 2008). A growing body of economic analysis suggests that forest transitions are linked to broader processes of land-use change in a particular country or region, and these are frequently shaped by market and policy processes outside the forestry sector per se (Mather, 2007; Barbier *et al.*, 2010). As Barbier *et al.* (2010, p. 99) explained: '[F]orest cover changes over time as the value of one land use relative to the value of its competing use changes over time...[I]f in any time period, the marginal benefits from agricultural production exceed the forest benefits, more land will be allocated to agricultural conversion.'

Although the immediate drivers of deforestation will vary from one context to another, clearance of forests is often encouraged by policy, market and institutional factors that increase the relative profitability of expanding croplands and pasture, versus maintaining forest cover on the one hand, versus intensifying existing agricultural lands on the other (Angelsen and Kaimowitz, 1999; Chomitz *et al.*, 2007).[6] While decisions concerning forest conversion are heavily shaped by the economic returns expected from competing land-use options, such decisions are rarely determined by market forces alone and, indeed, are often guided by distorted economic and policy incentives. As Barbier *et al.* (2010, p. 99) argued, '[T]he actual values that are used to allocate land may be far from optimal; [and] the presence of market, policy, and institutional failures... can distort economic and political incentives [and] lead to bias in favour of one type of land use over the other.' Market and governance failures encouraging tropical forest conversion and fragmentation commonly include weak property rights, inappropriate taxes and subsidies, and a general tendency to ignore the range of non-market environmental benefits that primary forests (even after logging) provide.

In recent years, the conversion of tropical forest lands to agriculture has increased dramatically in some regions with the leasing and sale of arable land to foreign companies and governments. These large-scale land acquisitions and commercial transactions, often termed 'land grabs', are undertaken by international corporations, sovereign wealth funds, foreign governments, private equity firms, and domestic corporate and state actors (Borras and Franco, 2012; Seo and Rodriguez, 2012). The objective on the part of governments is to secure food and fuel for growing populations in countries with poor agricultural land, or in those suffering from depleted soils, water shortages and projected energy deficits.[7] Much of the 'grabbing' of arable land has occurred in Africa – where land is comparatively cheap and considered plentiful and underutilized – as well as in South-east Asia, Latin America and Eastern Europe (McMichael, 2011). In many cases, large land transactions are linked to broader trade and investment agreements between resource-rich and investor country governments. In this way, sustainability concerns are often subordinated to the 'realpolitik' of commercial and diplomatic interests.

Throughout the tropics, roads are a key determinant of land use, and there is often a strong correlation between the development of roads and forest fragmentation (Angelsen and Kaimowitz, 1999; Chomitz *et al.*, 2007). On the one hand, roads make forest landscapes more accessible, facilitating the movement into these areas of agricultural settlers and large-scale commercial actors. On the other hand, roads link forest–agricultural frontiers to markets, thereby raising the profitability of food crops and agro-industrial commodities produced in these regions. Road networks play an integral role in shaping the geographical distribution of particular types of land use, and in this way they significantly influence the spatial geometry of forest fragmentation (Perz *et al.*, 2008a). The specific impacts of roads on forests are shaped by a host of 'road governance' issues, including: who builds and operates the roads and who has access; how decisions are made concerning routing, distance, and function; whether roads are paved or unpaved; how roads link to other infrastructure (e.g. ports, markets and other roads); and how road projects are financed (Perz *et al.*, 2008a,b). Analyses of the effects of roads on forest fragmentation in Amazonia and elsewhere have distinguished between the impacts of 'official roads' built by government agencies and 'unofficial roads' built by non-state actors (Box 10.3).

Globalization and the Governance of Transnational Commodity Networks

Over the past quarter of a century, the transformation of tropical forest landscapes has been driven, to a significant extent, by the combined effects of globalization (Lambin and Meyfroidt, 2011). Global markets for forest products and agricultural commodities have both expanded and become more integrated with the removal of tariff and non-tariff barriers to trade under GATT, the World Trade Organization, and other multilateral trade agreements. At the same time, liberalization of the global financial sector has facilitated transnational flows of capital, technological innovations have enabled the

Box 10.3. Roads and forest fragmentation in the Brazilian Amazon

In Amazonia, the construction of roads has been a major determinant of land-use change and has heavily influenced patterns of forest fragmentation. Official roads include inter-regional highways built or financed by governments and multilateral institutions, and their effects on deforestation and fragmentation are often macro in scope. They often 'form sparse networks, leaving large blocs of forest intact since parallel corridors are often hundreds of kilometres apart' (Perz *et al.*, 2008b, p. 1891). However, by connecting agricultural frontiers with ports and markets, official roads have been instrumental in facilitating the expansion of Brazil's cattle and soy industry into forested regions. One well-documented example is the BR-163 highway, which has functioned as a soy corridor running north to south through Pará and Mato Grosso states (Nepstad *et al.*, 2002; Fearnside, 2007). A significant factor shaping the impacts of roads on forests is whether they are surfaced, as surfacing substantially reduces transport costs, which, in turn, increase the profitability of land-extensive investments in frontier regions. The development of surfaced roads – and in some cases, even the expectation of road paving – causes land values to rise, and often encourages forest clearing by speculators as a way to establish land claims (Fearnside, 2007).

Unofficial roads, by contrast, are built by non-state social actors and are specifically instrumental in converting forests to other land uses. According to Perz *et al.* (2008b, p. 1891), '[U]nofficial roads form much denser networks in landscapes because they are often spaced only a few kilometres apart...[and these] networks fragment forest cover into smaller, more irregularly shaped and often more isolated patches that are more ecologically vulnerable.' In many cases, unofficial road networks extend outwards from much larger official roads, while in other places, they emerge where forests have initially been opened up by logging roads. The specific effects that unofficial roads have on the geometry of forest fragmentation are often shaped by local biophysical conditions as well as the land-use and colonization histories through which they are constructed. Perz *et al.* (2008a, p. 87) described the fragmentation effects of a road network where colonization was planned by INCRA, the national land settlement agency, as having a '"fishbone" pattern, characterized by elongated, irregularly shaped fragments organized in more or less parallel rows between deforested areas.' By contrast, in areas where roads have been developed 'spontaneously' by land users, the sizes of forest fragments vary with distance from towns: 'Close to towns, most land has been cleared, leaving isolated fragments, whereas far from towns, there are relatively few, large, rectangular clearings surrounded by large blocks of forests' (Perz *et al.*, 2008b, p. 90).

rapid exchange of information on a global scale, and infrastructure and logistics development has led to increased efficiency in the transport of goods worldwide. The effects of these changes on tropical forest landscapes have been powerfully reinforced by the simultaneous rise of China – and to a lesser extent, India and other emerging economies – as major importers and, in some cases, processors and re-exporters of forest products and agricultural commodities.

Increasingly, large-resource firms are sourcing raw materials from multiple geographical locations and coordinating these into centralized and highly capitalized production facilities (e.g. agro-industrial complexes, biofuel refineries, large pulp and paper mills) for processing and onward export (Bridge, 2008). These global production networks are establishing new divisions of labour, accessing financial markets in new ways and encouraging modes of governance based upon increasingly rapid turnover of capital – collectively leading to a more intensive commodification of tropical forest landscapes and the resources they contain (Dauvergne and Lister, 2011). Examples of global production networks in the forestry sector include the transnational wood fibre supply chains that supply Japanese and Chinese pulp and paper mills, which in turn generate processed paper products for domestic markets and for export to the USA and Europe, or IKEA's global wood furniture production system, in which plantation timber is sourced from Brazil, processed in Vietnam and then re-exported to consumers in the EU. Where such commodity networks are documented and transparent, there are opportunities for implementing improved sustainability regulations that might serve to limit associated forest fragmentation effects.

However, in cases where such global supply networks are non-transparent and/or illegal, there are significant barriers to the sustainable management of forest lands.

Different forest zones and global forest-based commodity supply chains can be governed through institutions at multiple scales. These include, for instance: traditional state authorities and national legal frameworks; corporate social responsibility programmes, commodity-specific roundtables (e.g. the Roundtable on Sustainable Palm Oil), and other voluntary standards; global financial mechanisms, such as the Equator Principles; consumer/market-based regulation (e.g. eco-certification with third-party auditing); or new legal-verification import requirements such as through the US Lacey Act Amendment or the EU FLEGT programme. In practice, however, the effectiveness of these regulatory systems is often uncertain. The rapid expansion of transnational governance requirements into developing country resource sectors and local contexts raises not only issues over the effectiveness and public accountability of market-based governance systems but also issues of developing country sovereignty (Vandergeest and Unno, 2012). At times, private-sector actors have voiced opposition to so-called 'green tape' and the increasing transaction costs involved in meeting proliferating certification standards.

Alternatively, some transnational corporations manage forest-land assets through large concessions that seemingly operate on the blurred edges of formal, legal regulation (e.g. Global Witness, 2013, in Laos and Cambodia). In such fragile and locally weak state environments, there can be significant challenges for forestry authorities to monitor and regulate effectively powerful transnational actors, who operate with high-level political backing. At the level of central state governance, a number of forest-rich, authoritarian regimes in South-east Asia and elsewhere may be undergoing a process of 'illiberal adaptation' (Felker, 2008), establishing new accommodations with neoliberal economic forces, while also working to consolidate the state's internal power and authority and forestry patronage networks. The prospects for stemming the tide of forest fragmentation and deforestation in such contexts would appear to be less positive.

Viewed from another perspective, the specific challenges associated with managing fragmented forests have led to the development of new approaches to forest management and conservation. The past decade, for instance, has seen an emerging consensus among corporate actors, conservation organizations and professional forestry institutions on the importance for forestry planning and management decisions to take account of the broader landscapes with which (fragmented) forests are situated (International Tropical Timber Organization, 2005; Sayer and Maginnis, 2007). In particular, the concept of landscape mosaics explicitly recognizes the fact that forest fragments often exist in proximity to lands being managed for a variety of other uses; and even in fragmented form, forest resources can provide essential economic, socio-cultural and environmental functions at the landscape level. This approach further encourages collaborative resource management practices among the often-diverse sets of stakeholders whose management decisions are likely to affect forests within a given landscape.

Among conservation organizations, the fragmentation of tropical forests has significantly influenced the strategies, institutional arrangements and international norms commonly used for biodiversity preservation. Increasingly, conservation efforts are now focusing on the preservation of not only large, intact forests but also forest islands containing vulnerable flora or fauna and/or unique ecosystems within landscapes undergoing transformation. This approach has led many of the world's largest conservation organizations – including the Worldwide Fund for Nature, The Nature Conservancy and Greenpeace, among many others – to collaborate with corporate actors directly involved in large-scale conversion of tropical forests, in an effort to protect fragments of natural forest within areas being converted to other land uses. Where fragmented forests are conserved to protect priority species, the size of the fragments and proximity to other fragments are frequently points of critical concern. These can determine whether the forested areas being protected are of sufficient size to sustain viable populations of key species. The use of biodiversity corridors to link forest fragments can also facilitate the movement of migratory species and/or those requiring a sizeable range, as well as to maintain a more diverse pool of genetic resources than a single fragment could provide.

In some cases, stakeholders have voiced concerns that corporate actors may be using such collaborations with conservation organizations – and, paradoxically, some of the tools designed to conserve forest fragments within multi-use landscapes – to legitimize continued clearing of natural forests. The use of 'high conservation value forest' (HCVF) assessments by Indonesia's leading pulp and paper producers, for instance, has raised concerns that a tool designed to protect priority conservation areas may in fact be giving the companies a 'green light' for large-scale forest conversion (Edwards *et al.*, 2012) (Box 10.4).

Displacement of Customary Management Systems and Socio-cultural Fragmentation

In many tropical countries, the assertion of state control over forests has been accompanied by the displacement of forest management systems utilized by local communities. In particular, national forestry laws have frequently endowed state agencies with far-reaching administrative control over forest resources, while placing constraints on customary tenure institutions (Colchester, 2006; Sunderlin *et al.*, 2008). For indigenous groups and other rural communities, this has often meant that access to land and forest resources – commonly including areas they had owned and managed for generations – has become restricted and many customary management practices deemed illegal. The resulting legal ambiguities and widespread

Box 10.4. Mega-scale pulp mills and forest fragmentation in Indonesia's Riau province

Since the mid-1990s, Indonesian-based multinationals Asia Pulp & Paper (APP) and Asia Pacific Resources International Ltd (APRIL) have operated two of the world's largest kraft pulp mills in Riau province on the island of Sumatra. With a combined pulp production capacity of 5.4 million air-dried tonnes per year (in 2013), the two producers' flagship mills – APP's Indah Kiat and APRIL's Riau Andalan Pulp & Paper (Riaupulp) – consume approximately 24 million tonnes of wood fibre annually. For the past two decades, this structural demand for wood fibre has been a significant driver of forest fragmentation in Riau and beyond (Barr, 2001b; Obidzinski and Dermawan, 2012). Forestry companies affiliated with APP and APRIL have converted large areas of degraded natural forest to fast-growing pulpwood plantations under the Ministry of Forestry's Hutan Tanaman Industri (HTI) industrial plantation policy. Moreover, the pulp mills' demand for fibre has encouraged the clearing of forested areas for the development of oil palm estates and other land uses. A 2008 study by the Worldwide Fund for Nature found that Sumatra lost 49% of its natural forest cover between 1985 and 2008/2009, and the losses were heaviest in Riau where Indah Kiat and Riaupulp are located (Worldwide Fund for Nature, 2008).

In response to sharp criticisms from civil society organizations and growing concerns from pulp and paper buyers, both APP and APRIL have implemented a range of initiatives aimed at persuading stakeholders that their operations are sustainable. Prominent among these has been a commitment from both companies to refrain from clearing areas designated as HCVF. Under these commitments, APP and APRIL have agreed to conduct HCVF assessments of all forests within their concessions prior to the clearing and conversion of these areas to pulpwood plantations. Proponents of this approach view HCVF assessments as an essential tool for conservation, which can effectively protect fragments of natural forest that have important ecological or social functions. More sceptical stakeholders are concerned, however, that, in Riau's highly industrialized landscape, APP and APRIL may be using HCVF assessments to create the aura of sustainability and technical legitimacy for the continued clearing of degraded natural forests on a large scale (Edwards *et al.*, 2012). Moreover, significant questions remain about the functional viability of forest fragments designated as having a 'high conservation value', as neither company routinely discloses where these areas are located or how they are managed after the HCVF assessment has been conducted.

conflicts represent a major breakdown in forest governance, and, in many regions, these have contributed to forest degradation and fragmentation and to negative impacts on the well-being of forest-reliant communities.

The subordination of customary tenure and management systems to state forestry laws was often initially imposed during the colonial era but has since been reinforced and expanded by post-colonial states across the tropics.[8] In many countries, state forestry agencies have used their authority not only to promote commercial development within the sector but also to restrict customary forest management and land-use practices. In both insular and mainland South-east Asia, for instance, forestry departments have adopted policies aimed at eradicating swidden agricultural systems in upland areas. As Fox *et al.* (2009, p. 308) explained, 'Post-colonial forest management authorities in these countries oversaw the massive expansion of national production and protection forests, the implementation of fast-growing timber and pulp plantations, development of industrial forestry policies, and in concert with these new land uses, the widespread criminalization of swidden cultivation.'

In some cases, governments have forcibly removed swidden communities from upland forest areas and resettled them in sites designated for sedentary agriculture (Dove, 1983; Baird and Shoemaker, 2007; Padoch *et al.*, 2007). Ironically, state forestry agencies have often justified such policies as being necessary to protect (state-controlled) forests from being fragmented by customary agricultural practices. A substantial body of research has determined, however, that, in the absence of severe population pressures or enclosure due to 'artificial' land use squeezes (such as the zoning of exclusionary protected or conservation forest areas, which confine swiddeners into increasingly smaller areas), long-fallow swidden agricultural systems can be highly appropriate and broadly sustainable strategies for producing food and managing forests sustainably in upland ecosystems (Dove, 1983, 1993; Fox *et al.*, 2000; Padoch *et al.*, 2007). This is not to suggest that community-managed forests are universally sustainable, and many communities have moved into management practices that are as likely to fragment forests as those promoted by state forestry agencies (cf. Garcia *et al.*, 2010; Feintrenie *et al.*, 2010; Temudo, 2012). The point, rather, is that traditional management systems continue to be marginalized and criminalized today without regard to the extent to which they are sustainable or not.

With the creation of the political forest, the state's view of what constitutes a forest was often superimposed over deeply-rooted social and cultural landscapes. Complex traditional forest management systems were undermined and, in many instances, forcibly circumscribed. Forest fragmentation can therefore be understood as both a biophysical process – i.e. in terms of its impacts on forest spatial structure and ecosystem functioning – and a socio-cultural process. In many cases, policies aimed at curtailing customary agricultural and forest management practices have effectively fragmented socio-cultural landscapes within which forests had been managed sustainably for centuries, with enormous implications for the health, well-being and cultural resilience of these groups (e.g. Dove, 1993; Balee, 1994; Posey, 2000; Fox *et al.*, 2009). Squeezed into increasingly small and marginal lands, and prohibited from accessing forests and resources, upland communities in many areas have been set up for unsustainable livelihoods, and, as discussed below, the pressure on their land and resources continues.

Vested with far-reaching legal authority to administer forests, state forestry agencies have often viewed the territory under their control as being essentially devoid of human occupants and unencumbered by competing tenurial claims. Within these agencies, government planners have routinely drawn land-use maps and demarcated the boundaries of timber concessions and agro-industrial estates with little attention to the locations of villages or local agricultural and forest management systems. Sociologist James Scott (1998) described this process as 'seeing like a state', noting that what state institutions are able to see is generally guided by the policies and plans they are charged with implementing, rather than the physical or socio-cultural realities on the ground.[9] Across the tropics, however, a large yet undetermined number of people – minimally, tens of millions and perhaps hundreds of millions[10] – live within and around the boundaries of state-controlled forests (World Bank, 2002; Colchester, 2006). In Indonesia, for example, the Ministry of Forestry has reported that some 40,800 villages are located in and around the national forest estate, and 25% of these are either situated within or immediately adjacent to the boundaries of the forest estate (Departemen Kehutanan, 2009). As these figures suggest, significant areas of state-controlled forestlands are (still) actively managed by local communities under various agroforestry systems (Fay and Sirait, 2004; Contreras-Hermosilla and Fay, 2005).

In many contexts, the state's failure to 'see' forest-reliant communities – much less recognize their rights to forest land and resources – has resulted in tenure disputes and, at times, violent conflicts. This has commonly occurred when state forestry agencies have allocated timber concessions or plantation licences for areas that overlap with territory claimed as customary domain by local communities (Harwell, 2003). Where formal institutions for resolving such disputes are weak or non-existent, resistance strategies by communities have frequently included the blocking of logging roads, the seizure of heavy equipment, the burning of forestry offices and the destruction of timber camps or plantation sites. Operating within such areas, timber and plantation companies have often drawn on government security forces and/or private security contractors to protect their activities from disruption – and in many cases, this has effectively extended the militarization of commercial forestry within such landscapes (Harwell, 2010).

Whether or not disputes over forest tenure ultimately lead to violence, the potential for conflict serves as a strong disincentive for all stakeholders – companies, communities and government agencies – to make significant investments in the sustainable management of forests over the long term. In contrast, competing tenure claims and high levels of uncertainty over how these will be resolved essentially encourage all parties to pursue the maximum benefits they can capture from forests in the short term. As tenure rights and control over forests become increasingly contested, forests are subject to opportunistic expropriation by many different actors, leading to a 'tragedy of open access' and large areas of 'ambiguous lands' (Hardin, 1968; Sato, 2000). Collectively, these pressures frequently lead to uncontrolled logging and conversion of forests at multiple scales, and over time, this can result in both the loss of forest cover and the degradation and fragmentation of forests that remain standing.[11]

Such outcomes have become especially prevalent in areas affected by 'land grabbing', as described above. Many land deals have resulted in the displacement of local communities, who often lack formal property titles and use their land under customary tenure arrangements. Deals between foreign investors and local governments are often arranged outside the public sphere, with very little transparency and external accountability, and so communities may not even be part of the negotiations in which their land is sold or leased (Cotula *et al.*, 2009; Seo and Rodriguez, 2012).[12] As Zoomers (2011, p. 18) noted:

> The poorer groups are usually the first to lose ground, certainly if they still do not possess formal property rights despite large-scale land titling programmes (people with only customary rights enjoy little protection from the law). At the same time, it is mainly the poorer groups who first decide to sell their land, because of their need for capital (distress sales), and new means of livelihood are not always readily available.

Over the past two decades, efforts have been made to reverse some of the damage resulting from state domination of the forest territories, and the denial of customary rights and practices. Significantly, there has been a trend towards greater recognition of the land, resource, human and cultural rights of indigenous peoples in international and national law (e.g. the UN Declaration on the Rights of Indigenous Peoples, 2007; the Convention on Biological Diversity, 1992 (e.g. Article 8j); the International Labor Organization's Convention No. 169; Posey, 2000). Among corporate actors involved in commercial forestry, agribusiness, and extractive industries, there has also been a growing adoption of principles of 'free, prior, and informed consent' as best practice in engaging with local communities likely to be affected by the operations of these companies (Colchester and Ferrari, 2007).

At the same time, many countries have moved to decentralize administration of forests within governments and to devolve control over forests from governments to local people (Ribot, 2002; Sunderlin *et al.*, 2008; Moeliono *et al.*, 2009). Decentralization initiatives have had mixed results in terms of impacts on forests and rural communities, and the transfer of authority to local groups has often resulted in management practices that are no more sustainable than those administered by national governments (see Box 10.5). In some tropical countries, moreover, national governments have aggressively sought to recentralize administrative control over forests (Ribot *et al.*, 2006). Several tropical forest countries, mostly in South America, have also made significant changes in forest tenure in the last ten years.[13] Brazil, Peru and Bolivia, for instance, have significantly increased land transfers to indigenous and community control. However, the granting of rights has not guaranteed their realization: in these countries, oil, gas and mining exploration has proceeded in violation of indigenous land titles, and in some cases, loggers, ranchers and miners have moved into extractive reserves with limited government response (Sunderlin *et al.*, 2008).

Conclusion

In this chapter, we have examined the links between commercial logging and forest conversion, governance failures and the fragmentation of tropical forests. We have shown how, as with broader processes of deforestation and degradation, forest fragmentation

Box 10.5. Forest fragmentation through conversion of 'degraded' community-managed forests into plantations in Laos

In Laos, regional and transnational companies have sought large-scale land concessions from the Lao state for rubber and eucalyptus plantation projects. While the Lao state claims that there are thousands of hectares of available 'degraded' forest land available in each province for allocation to companies, in reality almost all accessible rural forest lands in Laos are claimed and managed under customary community ownership (Barney, 2008). Degraded forest lands are zoned and demarcated out of complex, community-managed forest landscapes, and allocated to companies through the state Participatory Land Use Planning (PLUP) policy. Typically, these degraded, 'unused' forests are community-managed swidden fallow lands, and are very important for local agricultural and food security systems and sociocultural identities.

In Laos, some plantation companies ignore the formal land policy framework and, seemingly with high-level political backing, engage in blatant seizures of land from communities. Other firms, whose international commitments and reputations hold them more closely to formal legal procedures, have found the promised plantation concession land impossible to secure in practice. They have been required to undertake detailed and time-consuming negotiations with communities in their allocated project areas in an attempt to secure plantation land. With the enforced limitations on expressions of dissent against state policy in Laos, simmering land conflicts between companies and communities in Laos are now prevalent (e.g. Vientiane Times, 2013).

In mid-2012, the Lao Government announced a freeze on concessions for eucalyptus and rubber until 2016, to allow time to review the effectiveness of state policies in the plantation sector. As with a series of previous plantation concession moratoria between 2007 and 2009, however, there are questions concerning the ability of the central agencies to enforce these policy decisions effectively, due to the extent of discretionary decision making that characterizes the sector. In this way, the politics of land ownership and tenure policies in developing countries, the role of multiple state actors, institutions and bureaucracies, diverging corporate norms and regulatory standards, the high global demand for forest-land commodities, and the contested politics of state–society relations can all be intricately bound up with the spatial patterns of tropical forest fragmentation.

is often driven by weaknesses, contradictions and breakdown of forest governance institutions. The bases for forest governance failures include: (i) poorly formulated policies and regulations, especially related to commercial timber extraction; (ii) corruption and weak law enforcement; (iii) the influence of patronage politics and the dominance of commercial forestry by political and military elites; and (iv) conflicts between national forestry laws and customary tenure regimes.

However significant the effects of forest governance failures may be, forest fragmentation in many parts of the tropics is also driven by extra-sectoral governance processes, which provide strong incentives for forest cover removal. Most visibly, state-led spatial planning processes often promote the conversion of forested areas to a variety of other land uses, including agro-industrial tree crops, ranching, mining, smallholder agriculture, roads, and other infrastructure and development projects. One significant effect of land-use planning processes is that the spatial distribution of forest fragmentation is frequently non-random and is heavily shaped instead by government policies and planning decisions.

Changes in forest land use are also linked to increasingly globalized financial flows and the extension and deepening of global commodity markets, as well as the recent emergence of land grabbing linked to a new model of state capitalism and sovereign wealth funds. In this way, government trade and investment policies, as well as macroeconomic and fiscal policies, also represent important drivers of deforestation and forest fragmentation. Although such policies are generally designed to affect either the economy as a whole or a range of sectors well beyond forestry, they often have profound impacts in shaping the rate, extent and locus of forest cover loss and fragmentation.

The links between governance and forest fragmentation in the tropics are rarely unilinear, however. As forests become increasingly fragmented at multiple scales, forest fragmentation itself has emerged as a problem that requires innovative governance solutions. As state-controlled forest zones become increasingly fragmented, for instance, there is a growing recognition that effective management and protection of remaining forests will require meaningful forest tenure reform to strengthen rights of rural communities, restrictions on which often have their roots in colonial forest policies (Sunderlin *et al.*, 2008). For government forestry planners and commercial land users, there is also a growing recognition of the need to manage fragmented forests as part of the broader landscapes within which they are situated (Sayer and Maginnis, 2007). These commonly include a mosaic of different land uses managed by a variety of actors.

For conservation organizations, the challenges of protecting fragmented forest ecosystems highlight a need for innovative tools and institutional mechanisms. These are needed on the one hand to ensure that the areas conserved can support viable populations of critical species in spite of being spatially dispersed, and on the other hand to incorporate biodiversity conservation objectives into the management of broader landscapes, including those managed by corporate actors and by local communities. Finally, for proponents of REDD+, forest fragmentation poses significant challenges both for securing adequate scale of forest carbon projects and for managing, measuring, reporting and verification processes across fragmented landscapes without undue transaction costs.

Notes

[1]Prominent institutions that have promoted sustainable forest management through selective logging timber concessions include, among others, the Forest Stewardship Council, International Tropical Timber Organization, United States Agency for International Development, International Finance Corporation, Center for International Forestry Research, World Wide Fund for Nature, The Nature Conservancy and Rainforest Alliance.

[2]Rice *et al.* (1997, p. 46) noted, for instance, that 'real interest rates on dollar-denominated accounts in Bolivia...averaged 17 per cent [during the mid-1990s], compared with 4 per cent in the US. Similar high rates of interest are common in most Latin American countries.'

[3]Significantly, governance failures related to rent seizing are not limited to the allocation of timber concessions to state elites. Rather, they can also result from corruption and misuse of forestry revenues after they have been collected by the government. The loss of several billion dollars from Indonesia's Reforestation Fund, a government-managed fund established in the late 1980s with a mandate to finance reforestation and forest rehabilitation, provides a striking example of this (Barr *et al.*, 2010).

[4]As of March 2013, the EU had either signed or was in the process of negotiating VPAs with the following countries: Ghana, Cameroon, Liberia, Central African Republic, Republic of Congo, Democratic Republic of Congo, Gabon, Côte d'Ivoire, Indonesia, Malaysia, Vietnam, Guyana and Honduras. A review published in 2010 found that the VPA process was motivating participating producer countries to improve the monitoring and tracking of timber produced in their jurisdictions (Lawson and MacFaul, 2010). The challenges they face, however, are reflected in the following characterization of Cameroon's timber inspection system: 'The system does not track back to the stump – logs are often first marked many kilometres from the location of harvesting. The system also does not allow for easy reconciliation, and export volumes are rarely checked against licensed production volumes. Although the transport documents include some tamper-proofing measures, they are not sufficient to prevent counterfeiting. More timber transportation permits are often issued to logging companies than are required, and unused permits are not recovered; this has allowed companies to launder illegal logs into the system. The situation is showing signs of improvement, however. A new and more sophisticated timber-tracking system...is currently being developed in the context of the VPA being negotiated with the EU' (Lawson and MacFaul, 2010, p. 23).

[5]It should be noted that the conversion of natural forest to pulpwood plantations is generally not reflected in official deforestation statistics (including, for instance, those published by the FAO or Indonesia's Ministry of Forestry). The FAO defines 'deforestation' as 'the conversion of forest to another land use or the long-term reduction of the tree canopy below the minimum 10 per cent threshold' (Schoene *et al.*, 2007). Under this definition, the replacement of natural forest by monoculture plantations of exotic tree species is not considered to be deforestation, as it is not considered to be a change of land use or a reduction of tree canopy below 10%. Nevertheless, the development of pulpwood plantations in areas previously covered by natural forests can have many of the same fragmentation effects in forest landscapes as would occur under conversion to oil palm or other non-forestry land uses (Lang, 2009).

[6]According to Angelsen and Kaimowitz (1999, p. 81), higher prices for agricultural goods produced in frontier regions will typically stimulate increased forest conversion: 'As frontier agriculture becomes more profitable, both the existing population and migrants from other areas begin to shift resources into forest clearing. Higher prices also provide capital to put additional land under agricultural production.' Similarly, the availability of credit can lead to increased deforestation, but it generally does so only when it is allocated 'to finance activities associated with forest clearing, such as cattle ranching', as has been the case in Latin America (Angelsen and Kaimowitz, 1999, pp. 83–84). There is also strong evidence to suggest that 'higher rural wages reduce deforestation by making agricultural and forestry activities more costly', while 'at the individual household level, greater off-farm employment opportunities produce a similar effect by competing with such activities for labor' (Angelsen and Kaimowitz, 1999, p. 84).

[7]Significant momentum for exploitative land-grab processes has arisen from accelerating global commercial pressures on resources, rising commodity prices and the convergence of food, biofuel and wood-fibre markets (Roberts *et al.*, 2008). Extreme weather events, increased costs of petrochemical fertilizers and transportation, and competition from biofuels for agricultural land resulted in volatile food markets in 2007–2008. Spikes in food prices sent both investors and 'food-insecure' governments – like those of China (a net importer of food since the early 2000s), South Korea, Japan and the Gulf States – around the world in search of fertile agricultural land (Zoomers, 2011; Seo and Rodriguez, 2012). McMichael (2011, p. 12) described the ongoing land-grabbing processes as a response to a 'conjunctural crisis' in capitalist ecology: 'Rising food prices, peaking oil, emission mandates, and stalled investment funds find material resolution in the land grab, accompanied by an ideology of enclosure in the name of humanity (food) and the environment (green fuel).' Collectively, these forces have created intense new pressures for the conversion and fragmentation of tropical forests to supply food, biofuels and other commodities for expanding and geographically distant global markets.

[8]As Peluso and Vandergeest (2001) explained, colonial governments in South-east Asia placed restrictions on the unauthorized harvesting of valuable species such as teak, and introduced regulations limiting the activities of local people in the political forest. They granted exemptions for some customary practices

under the legal category Customary Rights, but many forms of customary resource use were taxed and required permits, or were considered 'forest crime'. In some cases, Customary Rights developed in one region were inappropriately applied to another (i.e. length of fallows allowed before land was considered abandoned and reverted to the state), and in other cases, communities were granted rights to undertake activities they did not practise in the first place. The limited recognition of Customary Rights helped governments to implement state forestry laws by striking a balance between consent and coercion, while appearing to be generous.

[9]Michael Dove argued that the inability (or unwillingness) of state actors to 'see' customary management practices has broadly undermined the effectiveness of government policies intended to promote both rural development and sustainable forestry: 'The main reason that policy fails (and will continue to fail) is that government officials have created a set of official myths about villagers' forest management which block their capacity either to see things from forest villagers' viewpoint or to formulate more equitable policies such as bottom-up management, community involvement in decision-making, or co-management strategies' (Dove 1993, cited in Peluso *et al.*, 1995, p. 212).

[10]The World Bank (2002) provides the following global estimates; however, these are not specific to the tropics: 'More than 1.6 billion people depend to varying degrees on forests for their livelihoods. About 60 million indigenous people are almost wholly dependent on forests. Some 350 million people who live in or adjacent to dense forests depend on them to a high degree for subsistence and income. In developing countries about 1.2 billion people rely on agroforestry farming systems that help to sustain agricultural productivity and generate income' (World Bank, 2002, cited by Colchester, 2006, p. 3).

[11]Barbier *et al.* (2010, p. 103) described the effects of tenure insecurity on farmers' land-use decisions in frontier regions from the perspective of resource economics: '[T]he lack of property rights and regulations means that farmers in forested regions face little or no opportunity costs of converting these forests to agriculture....Since the source of cleared land (i.e. forested areas) is an open access resource, farmers will keep converting land until the additional profits gained from clearing additional land just cover the labour and other costs involved in conversion. Perversely, however, insecure tenure and property rights mean that farmers are often required to make additional expenditures in defending and protecting their cleared land from appropriation by other farmers, squatters and even land speculators. These additional expenditures and effort may be prohibitive, and combined with the problem of declining agricultural yields on poor forest soils, often lead to abandonment of farmland after only a few years of initial cultivation. As a result, in many tropical forested regions, such as the Brazilian Amazon, excessive forest conversion to agriculture and substantial cropland abandonment occur simultaneously'.

[12]Foreign corporate and governmental entities claiming land in order to extract natural resources is, of course, not a new phenomenon, and the roots of current practices can be traced to the political forest created by colonial governments. As McMichael (2011, p. 1) noted: '[L]and grabbing is nothing new, and yet the recent "land rush" has its own distinctive features...land grabbing today continues the aggression of colonialism, but this time with a desperation born of crisis....'

[13]In a potentially significant development, Indonesia's Constitutional Court in May 2013 annulled the state's claims to customary forests. In response to a petition filed by Indonesia's National Indigenous Peoples' Alliance, AMAN (Aliansi Masyarakat Adat Nusantara), the Court eliminated the word 'state' from Article 11 of Indonesia's 1999 Law on Forestry, which previously stated that 'customary forests are state forests located in the areas of custom-based communities' (AMAN, 2013; Pasandaran, 2013).

References

AMAN (Aliansi Masyarakat Adat Nusantara) (2013) Constitutional Court Agrees on Judicial Review of UUK. May 16, 2013.

Angelsen, A. and Kaimowitz, D. (1999) Rethinking the causes of deforestation: lessons from economic models. *World Bank Observer* 14, 73–98.

Ascher, W. (1999) *Why Governments Waste Natural Resources: Policy Failures in Developing Countries*. Johns Hopkins University Press, Baltimore, Maryland.

Baird, I.G. and Shoemaker, B. (2007) Unsettling experience: internal resettlement and international aid agencies in Laos. *Development and Change* 38, 866–888.

Balee, W. (1994) *Footprints of the Forest: Ka-apor Ethnobotany – The Historical Ecology of Plant Utilization by an Amazonian People*. Columbia University Press, New York.

Barbier, E.B. (2005) *Natural Resources and Economic Development*. Cambridge University Press, Cambridge.

Barbier, E.B., Burgess, J.C. and Grainger, A. (2010) The forest transition: towards a more comprehensive theoretical framework. *Land Use Policy* 27, 98–107.

Barney, K. (2008) China and the production of forestlands in Laos: a political ecology of transnational enclosure. In: Nevins, J. and Peluso, N.L. (eds) *Taking Southeast Asia to Market: Commodities, Nature, and People in the Neoliberal Age*. Cornell University Press, Ithaca, New York, pp. 91–107.

Barr, C. (1998) Bob Hasan, the rise of Apkindo, and the shifting dynamics of control in Indonesia's timber sector. *Indonesia* 65, 1–36.

Barr, C. (2001a) Will HPH reform lead to sustainable forest management? Questioning the assumptions of the "sustainable logging" paradigm in Indonesia. In: Barr, C. (ed.) *Banking on Sustainability: Structural Adjustment and Forest Reform in Post-Soeharto Indonesia*. Center for International Forestry Research and WWF Macroeconomics Program Office, Washington, DC.

Barr, C. (2001b) Profits on paper: the political economy of fiber, finance, and debt in Indonesia's pulp and paper industry. In: Barr, C. (ed.) *Banking on Sustainability: Structural Adjustment and Forest Reform in Post-Soeharto Indonesia*. Center for International Forestry Research and WWF Macroeconomics Program Office, Washington, DC.

Barr, C. and Sayer, J. (2012) The political economy of reforestation and forest restoration in Asia-Pacific: critical issues for REDD+. *Biological Conservation* 154, 9–19.

Barr, C., Dermawan, A., Purnomo, H. and Komarudin, H. (2010) *Financial Governance and Indonesia's Reforestation Fund during the Soeharto and Post-Soeharto Periods, 1989–2009: A Political Economic Analysis of Lessons for REDD+*. Center for International Forestry Research (CIFOR), Bogor, Indonesia.

Bierregaard, R.O. Jr, Gascon, C., Lovejoy, T. and Mesquita, R. (eds) (2001) *Lessons from Amazonia: The Ecology and Conservation of a Fragmented Forest*. Yale University Press, New Haven, Connecticut.

Blundell, A.G. (2010) *Forests and Conflict: The Financial Flows that Fuel War*. World Bank, Washington, DC.

Borras, S. Jr and Franco, J. (2012) Global land grabbing and trajectories of agrarian change: a preliminary analysis. *Journal of Agrarian Change* 12, 34–59.

Boscolo, M. and Vincent, J.R. (2007) Area fees and logging tropical timber concessions. *Environmental and Development Economics* 12, 505–520.

Boucher, D.H., Elias, P., Lininger, K., May-Tobin, C., Roquemore, S. and Saxon, E. (2011) *The Root of the Problem: What's Driving Tropical Deforestation Today?* Union of Concerned Scientists, Cambridge, Massachusetts.

Brack, D. (2012) Excluding illegal timber and improving forest governance: the European Union's Forest Law Enforcement, Governance and Trade Initiative. In: Lujala, P. and Rustad, S. (eds) *High-Value Natural Resources and Peacebuilding*. Earthscan, London, pp. 211–220.

Bridge, G. (2008) Global production networks and the extractive sector: governing resource-based development. *Journal of Economic Geography* 8, 389–419.

Brown, D. (1999) *Addicted to Rent: Corporate and Spatial Distribution of Forest Resources in Indonesia: Implications for Forest Sustainability and Government Policy*. Report No. PFM/EC/99/06. Department for International Development, Jakarta.

Cerutti, P.O., Nasi, R. and Tacconi, L. (2008) Sustainable forest management in Cameroon needs more than approved forest management plans. *Ecology and Society* 13, 36.

Cerutti, P.O., Tacconi, L., Lescuyer, G. and Nasi, R. (2013) Cameroon's hidden harvest: commercial chainsaw logging, corruption, and livelihoods. *Society and Natural Resources* 26, 539–553.

Chomitz, K.M., Buys, P., De Luca, G., Thomas, T.S. and Wertz-Kanounnikoff, S. (2007) *At Loggerheads? Agricultural Expansion, Poverty Reduction, and Environment in Tropical Forests*. The World Bank, Washington, DC.

Clark, C.J., Poulsen, J.R., Malonga, R. and Elkan, P.W. Jr (2009) Logging concessions can extend the conservation estate for Central African tropical forests. *Conservation Biology* 23, 1281–1293.

Colchester, M. (2006) *Forest Peoples, Customary Use, and State Forests: The Case for Reform*. Paper presented to the 11th Biennial Congress of the International Association for the Study of Common Property (IASCP), Bali, Indonesia, 19–23 June. Forest Peoples Programme <http://dspace.cigilibrary.org/jspui/bitstream/123456789/12086/1/Forest%20Peoples%20Customary%20Use%20and%20State%20Forests%20the%20case%20for%20reform.pdf?1> (accessed 29 July 2014).

Colchester, M. and Ferrari, M.F. (2007) *Making FPIC – Free, Prior, and Informed Consent – Work: Challenges and Prospects for Indigenous Peoples*. FPIC Working Paper (June). Forest Peoples Programme, Moreton-in-Marsh, UK.

Contreras-Hermosilla, A. and Fay, C. (2005) *Strengthening Forest Management in Indonesia through Land Tenure Reform: Issues and Framework for Action*. Forest Trends, Washington, DC.

Convention on Biological Diversity (2014) Definitions. http://www.cbd.int/forest/definitions.shtml (accessed 9 July 2014).
Cotula, L., Vermeulen, S., Leonard, R. and Keely, J. (2009) *Land Grab or Development Opportunity? Agricultural Investments and International Land Deals in Africa*. UN Food and Agriculture Organization (FAO), Institute for Food and Agricultural Development (IFAD) and International Institute for Economic Development (IIED), London/Rome.
Dauvergne, P. and Lister, J. (2011) *Timber*. Polity Press, Cambridge.
Departemen Kehutanan (2009) Identifikasi Desa di Dalam dan di Sekitar Kawasan Hutan 2009. Ministry of Forestry, Jakarta.
Dove, M. (1983) Theories of swidden agriculture and the political economy of ignorance. *Agroforestry Systems* 1, 85–99.
Dove, M. (1993) Smallholder rubber and swidden agriculture in Borneo: a sustainable adaptation to the ecology and economy of the tropical forest. *Economic Botany* 47, 136–147.
Dwyer, M. (2013) Building the politics machine: tools for 'resolving' the global land grab. *Development and Change* 44, 309–333.
Edwards, D.P., Fisher, B. and Wilcove, D.S. (2012) High conservation value or high confusion value? Sustainable agriculture and biodiversity conservation in the Tropics. *Conservation Letters* 5, 20–27.
FAO (2011) *State of the World's Forests Report*. UN Food and Agriculture Organization (FAO), Rome.
Fay, C. and Sirait, M. (2004) *Indonesia's Agrarian and Forestry Legal Frameworks: Challenging the National Dual System of Land Administration*. Paper presented at the International Conference on Land Tenure, Jakarta, 11–13 October 2004.
Fearnside, P. (2007) Brazil's Cuiabá-Santarém (BR-163) highway: the environmental cost of paving a soybean corridor through the Amazon. *Environmental Management* 39, 601–614.
Feintrenie, L., Schwarze, S. and Levang, P. (2010) Are local people conservationists? Analysis of transition dynamics from agroforests to monoculture plantations in Indonesia. *Ecology and Society* 15, 37.
Felker, G. (2008) Southeast Asia and globalization: the political economy of illiberal adaptation. In: Kuhonta, E., Slater, D. and Vu, T. (eds) *Southeast Asia in Political Science: Theory, Region and Qualitative Analysis*. Stanford University Press, Stanford, California, pp. 274–301.
Fisher, B. (2010) African exception to drivers of deforestation. *Nature Geoscience* 3, 375–376.
Fox, J., Truong, D.M., Rambo, A.T., Tuyen, N.P., Cuc, L.T. and Leisz, S. (2000) Shifting cultivation: a new old paradigm for managing tropical forests. *Biosience* 50, 521–528.
Fox, J., Fujita, Y., Ngidang, D., Peluso, P., Potter, L., Sakuntaladewi, N., Sturgeon, J. and Thomas, D. (2009) Policies, political economy and swidden in Southeast Asia. *Human Ecology* 37, 305–322.
Garcia, C.A., Bhagwat, S.A., Ghazoul, J., Nath, C.D., Konerira, M.N., Cheppudira, K.G., Raghuramulu, Y., Nasi, R. and Vaast, P. (2010) Biodiversity conservation in agricultural landscapes: challenges and opportunities of coffee agroforests in the Western Ghats, India. *Conservation Biology* 24, 479–488.
Gibbs, H.K., Ruesch, A.S., Achard, F., Clayton, M.K., Holmgren, P., Ramankutty, N. and Foley, J.A. (2010) Tropical forests were the primary sources of new agricultural land in the 1980s and 1990s. *Proceedings of the National Academy of Sciences USA* 107, 16732–16737.
Global Witness (2009) *Undue Diligence: How Banks do Business with Corrupt Regimes* <http://www.globalwitness.org> (accessed 29 July 2014).
Global Witness (2013) *Rubber Barons: How Vietnamese Companies and International Financiers are Driving a Land Grabbing Crisis in Cambodia and Laos* <http://www.globalwitness.org/rubberbarons/> (accessed 29 July 2014).
Goncalves, P.M., Panjer, M., Greenberg, T.S. and Magrath, W.B. (2012) *Justice for Forests: Improving Criminal Justice Efforts to Combat Illegal Logging*. The World Bank, Washington, DC.
Hardin, G. (1968) The tragedy of the commons. *Science* 162, 1243–1248.
Harris, L.D. (1984) *The Fragmented Forest: Island Biogeography and the Preservation of Biotic Diversity*. University of Chicago Press, Chicago, Illinois.
Harwell, E. (2003) *Without Remedy: Human Rights Abuse and Indonesia's Pulp and Paper Industry*. Human Rights Watch 15, No. 1 <http://www.hrw.org/reports/2003/indon0103/Indon0103.pdf> (accessed 29 July 2014).
Harwell, E. (2010) *Forests in Fragile and Conflict-Affected States*. Program on Forests (PROFOR), Washington, DC.
Hodgdon, B. (2008) Frontier country: the political culture of logging and development on the periphery in Laos. *Kyoto Journal* 69, 58–65.
Hosonuma, N., Herold, M., De Sy, V., De Fries, R., Brockhaus, M., Verchot, L., Angelsen, A. and Romijn, E. (2012) An assessment of deforestation and forest degradation drivers in developing countries. *Environmental Research Letters* 7, 1–12.

Independent Evaluation Group (2012) *Managing Forest Resources for Sustainable Development: An Evaluation of World Bank Group Experience*. World Bank Group, Washington, DC.

International Tropical Timber Organization (2005) *Restoring Forest Landscapes: An Introduction to the Art and Science of Forest Landscape Restoration*. ITTO Technical Series No. 23. International Tropical Timber Organization/International Union for Conservation of Nature <http://www.itto.int/news_releases/id=10640000> (accessed 15 July 2014).

Irland, L.C. (2008) State failure, corruption, and warfare: challenges for forest policy. *Journal of Sustainable Forestry* 27, 189–223.

Kaimowitz, D. (2004) Conventional wisdom about sustainable forest management and a pro-poor forest agenda. In: Zarin, D.J., Alavalapati, J.R.R., Putz, F.E. and Schmink, M. (eds) *Working Forests in the Neotropics*. Columbia University Press, New York, pp. 379–387.

Karsenty, A. (2000) *Economic Instruments for Tropical Forests: The Congo Basin Case*. International Institute for Environment and Development, London.

Karsenty, A. and Gourlet-Fleury, S. (2006) Assessing sustainability of logging practices in the Congo Basin's managed forests: the issue of commercial species recovery. *Ecology and Society* 11, 26.

Karsenty, A. and Ongolo, S. (2011) Can 'fragile states' decide to reduce their deforestation? The inappropriate use of the theory of incentives with respect to the REDD mechanism. *Forest Policy and Economics* 18, 38–45.

Kenney-Lazar, M. (2010) *Land Concessions, Land Tenure, and Livelihood Change: Plantation Development in Attapeu Province, Southern Laos*. Faculty of Forestry, National University of Laos, Vientiane, Laos.

Lambin, E.F. and Meyfroidt, P. (2011) Global land use change, economic globalization, and the looming land scarcity. *Proceedings of the National Academy of Sciences USA* 108, 3465–3472.

Lang, C. (2009) Wilful ignorance: FAO and industrial tree plantations. *WRM Bulletin* 141 <http://wrm.org.uy/articles-from-the-wrm-bulletin/section4/wilful-ignorance-fao-and-industrial-tree-plantations/> (accessed 29 July 2014).

Laurance, W.F., Ferreira, L.V., Rankin-de Merona, J.M. and Laurance, S.G. (1998) Rain forest fragmentation and the dynamics of Amazonian tree communities. *Ecology* 79, 2032–2040.

Lawson, S. and MacFaul, L. (2010) *Illegal Logging and Related Trade: Indicators of the Global Response*. Chatham House, London.

Mather, A.S. (2007) Recent Asian forest transitions in relation to forest-transition theory. *International Forestry Review* 9, 491–501.

McElwee, P. (2004) You say illegal, I say legal: the relationship between 'illegal' logging and land tenure, poverty, and forest use rights in Vietnam. *Journal of Sustainable Forestry* 19, 97–135.

McMichael, P. (2011) The food regime in the land grab: articulating 'global ecology' and political economy. Paper presented at the International Conference on Global Land Grabbing, University of Sussex, 6–8 April.

Meijard, E. and Sheil, D. (2007) A logged forest in Borneo is better than none at all. *Nature* 446, 974.

Moeliono, M., Wollenberg, E. and Limberg, G. (eds) (2009) *The Decentralization of Forest Governance: Politics, Economics, and the Fight for Control of Forests in Indonesian Borneo*. London: Earthscan.

Nasi, R. and Frost, P.G.H. (2009) Sustainable forest management in the tropics: is everything in order but the patient still dying? *Ecology and Society* 14, 40.

Nepstad, D., McGrath, D., Alencar, A., Barros, A.C., Carvalho, G., Santili, M. and Vera Diaz, M.d.C. (2002) Frontier governance in Amazonia. *Science* 295, 629–631.

Obidzinski, K. and Dermawan, A. (2012) Pulp industry and environment in Indonesia: is there a sustainable future? *Regional Environmental Change* 12, 961–966.

OECD/DAC (2007) Principles for good international engagement in fragile states and situations. Organization for Economic Cooperation and Development (OECD), Paris.

Padoch, C., Coffey, K., Mertz, O., Leisz, S.J., Fox, J. and Wadley, R.J. (2007) The demise of swidden in Southeast Asia? Local realities and regional ambiguities. *Danish Journal of Geography* 107, 29–41.

Pasandaran, C. (2013) Constitutional court annuls government ownership of customary forests. *Jakarta Globe*, 17 May 2013.

Peluso, N.L. and Vandergeest, P. (2001) Genealogies of the political forest and customary rights in Indonesia, Malaysia, and Thailand. *Journal of Asian Studies* 60, 761–812.

Peluso, N.L., Vandergeest, P. and Potter, L. (1995) Social aspects of forestry in Southeast Asia: a review of postwar trends in the scholarly literature. *Journal of Southeast Asian Studies* 26, 176–218.

Perz, S., Caldas, M., Walker, R., Arima, E. and Souza, C. (2008a) Road networks and forest fragmentation in the Amazon: explanations for local differences with implications for conservation and development. *Journal of Latin American Geography* 7, 85–104.

Perz, S., Brilhante, S., Brown, F., Caldas, M., Ikeda, S., Mendoza, E., Overdevest, C., Reis, V., Reyes, J.F., Rojas, D., Schmink, M., Souza, C. and Walker, R. (2008b) Road building, land use and climate change: prospects for environmental governance in the Amazon. *Philosophical Transactions of The Royal Society B: Biological Sciences* 363, 1889–1895.

Posey, D.A. (ed.) (2000) *The Cultural and Spiritual Values of Biodiversity*. United Nations Environment Programme, Nairobi.

Repetto, R. and Gillis, M. (eds) (1988) *Public Policies and the Misuse of Forest Resources*. Cambridge University Press, New York.

Ribot, J.C. (2002) Democratic Decentralization of Natural Resources: Institutionalizing Popular Participation. Washington, DC: World Resources Institute.

Ribot, J.C., Agrawal, A. and Larson, A. (2006) Recentralizing While Decentralizing: How National Governments Reappropriate Forest Resources. *World Development* 34, 1864–1886.

Rice, R., Gullison, R. and Reid, J. (1997) Can Sustainable Management Save Tropical Forests? *Scientific American* 276, 44–49.

Roberts, D., White. A. and Nilsson, S. (2008) Convergence of food, fuel and fibre markets: driving change in the world's forests. *ArborVitae* 37, 8.

Ross, M.L. (2001) *Timber Booms and Institutional Breakdown in South East Asia*. Cambridge University Press, Cambridge, UK.

Rudel, T.K., DeFries, R., Asner, G.P. and Laurance, W.F. (2009) Changing drivers of deforestation and new opportunities for conservation. *Conservation Biology* 23, 1396–1405.

Sato, J. (2000) People in between: conversion and conservation of forest lands in Thailand. *Development and Change* 31, 155–177.

Sayer, J. and Maginnis, S. (2007) *Forests in Landscapes: Ecosystem Approaches to Sustainability*. Earthscan, London.

Schoene, D., Killman, W., von Lüpke, H. and Mette, M.L. (2007) *Definitional Issues Related to Reducing Emissions from Deforestation in Developing Countries*. Forests and Climate Change Working Paper 5, UN Food and Agriculture Organization, Rome.

Scott, J. (1998) *Seeing Like a State: How Certain Schemes to Improve the Human Condition have Failed*. Yale University Press. New Haven, Connecticut.

Seneca Creek Associates and Wood Resources International (2004) *"Illegal" Logging and Global Wood Markets: The Competitive Impacts on the U.S. Wood Products Industry*. American Forest and Paper Association, Washington, DC <http://www.illegal-logging.info/sites/default/files/uploads/1_AF_and_PA_summary.pdf> (accessed 29 July 2014).

Seo, K. and Rodriguez, N. (2012) Land grab, food security and climate change: a vicious circle in the global South. In: Chhetri, N. (ed.) *Human and Social Dimensions of Climate Change* <http://dx.doi.org/10.5772/50876> (accessed 29 July 2014).

Setiono, B. and Husein, Y. (2005) Fighting forest crime and promoting prudent banking for sustainable forest management: the anti-money laundering approach. CIFOR Occasional Paper No. 44. Center for International Forestry Research (CIFOR), Bogor, Indonesia.

Sist, P., Dykstra, D. and Fimbel, R. (1998) Reduced-impact logging guidelines for lowland and hill dipterocarp forests in Indonesia. CIFOR Occasional Paper No. 18. Center for International Forestry Research (CIFOR), Bogor, Indonesia.

Sunderlin, W., Hatcher, J. and Liddle, M. (2008) *From Exclusion to Ownership? Challenges and Opportunities in Advancing Forest Tenure Reform*. Rights and Resources Initiative, Washington, DC.

Temudo, M.P. (2012) The white men bought the forests: conservation and contestation in Guinea-Bissau, Western Africa. *Conservation and Society* 10, 354–366.

Tsing, A. (2004) *Friction: An Ethnography of Global Connection*. Princeton University Press, Princeton, New Jersey.

Vandergeest, P. and Peluso, N.L. (1995) Territorialization and state power in Thailand. *Theory and Society* 24, 385–426.

Vandergeest, P. and Peluso, N.L. (2006) Empires of forestry: professional forestry and state power in Southeast Asia, Part 2. *Environment and History* 12, 359–393.

Vandergeest, P. and Unno, A. (2012) A new extraterritoriality? Aquaculture certification, sovereignty and empire. *Political Geography* 31, 358–367.

Vientiane Times (2013) Land compensation tops parliament debate. *Vientiane Times*, 26 July 2013.

World Bank (1992) *World Development Report*. Oxford University Press, New York.

World Bank (2002) *A Revised Forest Strategy for the World Bank Group*. World Bank, Washington, DC.

Worldwide Fund for Nature (2008) Deforestation, forest degradation, biodiversity loss, and CO_2 emissions in Riau, Sumatra, Indonesia. *Worldwide Fund for Nature Indonesia Technical Report*, 27 February 2008.
Zimmerman, B. and Kormos, C. (2012) Prospects for sustainable logging in the tropical forests. *Bioscience* 62, 479–487.
Zoomers, A. (2011) Introduction: rushing for land: equitable and sustainable development in Africa, Asia and Latin America. *Development* 54, 12–20.

11 Future Forests: Fantasy or Façade? A Synthesis

Chris J. Kettle,[1]* Rhett Harrison[2] and Lian Pin Koh[3]
[1]ETH Zurich, Switzerland; [2]World Agroforestry Centre, Kunming, Yunnan Province, PR China; [3]University of Adelaide, Australia

Abstract

Reversing global forest fragmentation requires substantial changes in the way as a society we value and use forest land. We summarize here the challenges to reversing forest fragmentation and provide a synthesis of the major topics covered in this book. We follow this with an optimistic view of the many approaches that can be used to ensure that forest fragmentation is reduced with examples ranging from small-scale community ecological restoration in Scotland to large-scale ecological restoration in Sumatra. These inspirational projects provide hope for the future of global forest landscapes, but we still require changes in perception at multiple levels from individuals living in rural and still largely forested areas to global companies that incentivize large-scale land conversion. We hope that the pages of this book will go some way to ensuring the long-term persistence of forest landscapes that sustain a rich and fruitful habitat for generations to come.

Introduction

We face an enormous challenge to reverse the trend in global forest fragmentation. It is estimated that 30% of global forest land has been completely cleared and 20% has been degraded (WRI, 2011). The proportion of intact forest landscapes varies enormously across the globe, with the environmental and ecological implications of this varying from region to region. The majority of intact forest landscapes remain in the boreal forest and the humid tropics. Inevitably, global demand for food and fibre will increase as the human population climbs towards an estimated 9.6 billion in 2050 (Anon., 2013) further squeezing natural forests out of our landscapes.

It is unequivocal that global society depends upon functioning forest ecosystems, with the numerous ecosystem services they provide including sustaining biological diversity (at all its levels). Future forest landscapes must address the consequences of forest fragmentation if these are to be sustained. Despite enormous global importance, only a small fraction of forests (approx. 7.7%) fall within internationally recognized categories of protection (Schmitt *et al.*, 2009), with less than 10% of tropical forests protected (WWF, 2002). More than 50% of unprotected tropical moist forests have already lost half of their tree cover (Asner *et al.*, 2009). The natural recovery and assisted restoration of forest landscapes thus present a major global priority. Indeed, the

*chris.kettle@env.ethz.ch

ambitious Aichi Target 15 of the Convention on Biodiversity aims to restore 15% of degraded ecosystems in the coming 6 years (CBD, 2011), while the Bonn Challenge of the same year targets 150 million ha of land worldwide for reforestation (ICUN, 2011), estimated to have a value of US$85 billion per year to national and global economies. Even beginning to approach these targets is a multidisciplinary challenge that includes significant social, economic, ecological and technical hurdles. It will also require making difficult decisions and priority setting using limited economic and ecological resources. This book provides an overview of our understanding of global forest fragmentation in an effort to scientifically inform these challenges.

This final chapter aims to synthesize the major themes covered in the book and, using a range of case studies, to illustrate that multiple pathways exist to reverse forest fragmentation and increase the resilience of forest landscapes of the future. We do not aim to review the existing literature on forest restoration or conservation, as this has been amply covered in the literature (see for details: Holl and Aide, 2011; Lamb, 2011; Kettle, 2012; Stanturf *et al.*, 2012). Instead, by showcasing some exceptional examples of how different stakeholders have attempted to reverse forest fragmentation, we present an optimistic perspective of how functioning forest landscapes might be sustained in the future.

The Major Challenges to Reversing Global Forest Fragmentation

As illustrated by Bhagwat (Chapter 2, this volume), forest fragmentation can be both natural and anthropogenic in its origin. Human-induced fragmentation presents extremes in both the rate and spatial extent of forest fragmentation. The combination of this intensity has knock-on effects for the ecological quality of the intervening matrix, which is often more hostile in human-fragmented landscapes. The historical origins of many fragmented forest landscapes remain unclear. Advancing our understanding of these origins will provide important knowledge for prioritizing forest protection and restoration efforts. As highlighted by Laurance (Chapter 3, this volume), although the human drivers of forest fragmentation continue to depend on agricultural expansion, increasingly intensive resource management, plantations, mining and infrastructural development are exacerbating forest fragmentation. Global economic drivers pose major challenges to preventing further fragmentation, especially through large-scale industrial agribusiness. Engaging with these powerhouses of land-use change will be vital to reversing fragmentation across landscape scales (see Box 11.5).

If forest biodiversity is to persist in human-modified landscapes, then forest fragments and the intervening matrix must be sufficiently large and sufficiently permeable to sustain viable populations. The myriad of ways that taxa interact with these novel environments and with each other present enormous challenges for conservation scientists and managers alike. Moreover, as a society, we need to develop a broader tolerance for forest-dwelling animals in these landscapes, so that species may cross between fragments without persecution. Where wild animal populations are exploited, for meat or fur, this needs to be managed sustainably and with an understanding of the interactions between fragmentation and harvesting and population dynamics (Harrison *et al.*, 2013). As articulated by Döbert *et al.* (Chapter 4, this volume), the ability of species to survive in the face of human-induced landscape change will be more important than conservation efforts to shield species from it. Adopting a landscape approach involves managing highly complex systems with unpredictable responses and multiple competing interests. This necessitates a shift away from an entirely conservation-orientated focus towards a more holistic set of general principles to landscape management (Sayer *et al.*, 2013).

The quality and permeability of the intervening matrix within landscapes will be crucially important in determining how the genetic consequences of fragmentation manifest and the vulnerability of different forest taxa to fragmentation genetics (Chapter 5, this volume). Many of the negative genetic consequences or evolutionary responses to fragmentation are likely to have gone undetected for many species, and potentially crucial impacts probably still remain to be discovered. The synergistic interactions between fragmentation and defaunation are also likely to have major negative genetic consequences. Negative consequences of genetic fragmentation coupled with local habitat degradation can present a double blow for forest-tree recruitment (Ismail *et al.*, 2014). Novel genomic techniques that enable us to determine genes underlying specific traits or adaptations provide great opportunities for elucidating cryptic responses and supporting sustainable management of forest genetic resources at landscape scales. From the forest-tree perspective, this information is likely to be crucial for managing genetic resources in response to future climate change and mitigation strategies. Indeed, climate change mitigation strategies such as REDD+ (reducing emissions from deforestation and forest degradation) offer significant impetus for preventing and reversing forest fragmentation (Chapter 6, this volume). At the more local scale, this will help prevent negative impacts on local microclimates, such as increased drought events and fire frequency. Restoration of degraded forest patches, through enrichment planting or seeding with native tree species, and maintaining or enhancing native tree cover within the intervening matrix, will also help minimize these effects to local microclimates in human-modified landscapes.

One of the major debates concerning forest fragmentation is that of land sparing versus land sharing (Chapter 7, this volume). This debate centres on resolving conflicts between agricultural productivity and biodiversity conservation. Although the land-sparing debate has helped to crystallize our understanding of the challenges associated with multiple trade-offs, the landscape approach shifts away from this rather polarized perspective, and instead acknowledges that the potential benefits are rather case specific. Intensive agriculture will be both necessary and in certain contexts more efficient, while in other situations a wildlife-friendly agricultural system will provide great cultural richness, biodiversity benefits (and sometimes other ecosystem service benefits) and other societal advances, without significant costs to production. Identifying policy mechanisms to account for these multiple stakeholder interests will require a sound understanding of the economics of complex ecological landscapes, where the externalities associated with intensive production have been internalized. Once these difficult economic questions have been resolved more adequately, the cost–benefit analyses of alternative land-management decisions can be conducted to evaluate more fully the societal benefits. The costs of losing species, degrading ecosystem functioning and reducing ecosystem services must clearly be accounted for. As articulated by Klein *et al.* (Chapter 7, this volume), both land sparing and land sharing can provide biodiversity and crop-yield benefits. More importantly, the integration of a landscape approach that maximizes the utility of forest fragments for multiple uses is likely to enhance the livelihoods of local people and environmental benefits.

The implications of forest fragmentation for ecosystem function and the ability of future forests to sustain certain ecosystem services remain poorly resolved. For example, how does forest fragmentation influence the long-term ability of forest fragments to sequester carbon, maintain high tree species diversity or sustain yields of high-value timber species? Our current understanding of plant–animal interactions, which are likely to be especially vulnerable to forest fragmentation (Chapter 8, this volume), reiterates the importance of the intervening matrix quality. Equally, the synergistic effects of the surrounding matrix on forest function can be profound. The implications of these synergistic effects for the land-sparing approach, where intensive agricultural practices may be detrimental to the functioning of adjacent forest, cannot be underestimated. Studies are needed at landscape scales to explore these synergies. Indeed, although several experimental landscape-scale fragmentation studies have been established and have advanced our scientific understanding (e.g. the Stability of Altered Forest Ecosystems project in Asia, and the Biological Dynamics of Forest Fragments Project in the Brazilian Amazon), future studies should prioritize experimental approaches to investigate the effects of forest restoration or wildlife-friendly farming. For example, if an oil-palm company agreed to forgo 10% of its productivity for biodiversity benefits, what approach should we take? Would lower-density planting and understorey treatments to improve the permeability of the matrix reap higher rewards than restoring forest corridors? Advancing agroforestry practices through the conservation of traditional practices as well as identifying novel species and establishing new markets also has a considerable role to play in enhancing forest landscape permeability and ecosystem services.

Balancing these trade-offs between multiple objectives for land resources depends not only on economic and ecological drivers but also societal demands. These demands are influenced by increasing population growth and technological advances. As articulated by Garcia and Feintrenie (Chapter 9, this volume), the forest transition curve is a global phenomenon that underlies global forest fragmentation. It is the establishment of social values for biological diversity juxtapositioned with livelihood imperatives (see Boxes 11.2 and 11.3), material wealth and modern comfort levels, which ultimately drive local decisions. Enhancing the ability of currently highly forested countries to make the transition to developed nations without sacrificing their forest reserves and forest biodiversity is an international policy challenge. What we see in many nations with intact forest landscapes is a recapitulation of the transition curve tracked by many now-wealthy nations who have already denuded their forest landscapes (Corlett, 2013).

Approximately 75% of the world's forests are under the institutional control of government. Effective and equitable forest governance is a crucial component of reversing forest fragmentation. In many tropical developing countries, governance issues are especially acute because the great majority of forested lands are under state agency administration. As discussed by Barr *et al.* (Chapter 10, this volume), this poses both great opportunities and challenges, not least in establishing policies and guidelines to help achieve sustainable selective logging, improving law enforcement and reducing corruption. Understanding the socio-cultural implications of forest fragmentation is as important as the biophysical effects. To achieve resilient future forest landscapes will require balancing the needs of forest-dependent communities and novel market-based global forest governance initiatives. International treaties and agreements such as the United Nations Framework Convention on Climate Change and the Convention on Biological Diversity thus have significant potential for leveraging change in ratified countries.

Reversing Fragmentation through Drivers of Change

Rates of forest loss and fragmentation across the humid tropics have either remained high or shown only a slow decline. Equally, in the vast majority of developed countries, only a tiny proportion of the original forest remains. This underlies a vast potential for reversing forest fragmentation and increasing the proportion of forest in human-dominated landscapes. Indeed, the Global Partnership on Forest and Landscape Restoration (http://www.forestlandscaperestoration.org/) suggests that there are some 2 billion ha of deforested and degraded lands worldwide with opportunities for restoration, an area twice the size of the USA. This includes wide-scale landscape-level ecological restoration, mosaic restoration of forest patches within more complex landscapes and an additional 200 million ha of unpopulated land in the remote areas of the far north boreal forests. Agricultural croplands and urban areas represent an estimated additional billion hectares of former forest land, which could include isolated trees, hedgerows and planting along river banks. All of these activities will contribute to reducing the negative effects of global forest fragmentation. Restoration is burgeoning across the globe as evidenced by the growing number of databases such as that of the Society for Ecological Restoration (http://www.globalrestoration-network.org) and online resources through the Environmental Leadership and Training

Initiative (http://reforestation.elti.org/) to name but two. The recently established research coordination network of natural and social scientists, PARTNERS (http://partners-rcn.uconn.edu), aims to address the complexity of socio-ecological processes that influence forest restoration across the tropics.

Scaling up restoration to meet proposed targets will require substantial economic and social investment. The drivers for this investment will inevitably come from diverse sources and the following case studies illustrate both the challenges and power of individuals and communities to establish these drivers of change. It is perhaps optimistic to reflect on the fact that, globally, forest restoration and rewilding are already objectives in many nations. Despite natural forest originally covering 80% of the land area in the UK, by 1900 less than 5% remained. Although large-scale reforestation led by the state forest department began in the 1920s, less than 2% of land is covered by semi-natural woodland in the UK today. It is only in the last three to four decades that restoration of native woodland in the UK has become a significant activity. Box 11.1 illustrates how a rural community in Scotland has driven changes in forest cover through their own passion for the natural environment.

That we are seeing forest restoration occurring in developing countries of the humid tropics such as those highlighted below reflects hope that, despite not yet having completely denuded their forest landscapes to fuel their economic development, these countries may manage to retain a greater proportion of their natural forest than has been achieved in much of the temperate regions.

Historically, much impetus for forest conservation has been through conservation of individual charismatic megafauna, and this has also provided the motivation and resources for habitat restoration of flagship species such as the Sumatran orang-utan described in Box 11.2. Such projects have a key role to play not only in reversing forest fragmentation but also in raising awareness and education of local communities to help further reduce forest degradation.

Increasingly we are aware of how important social factors are in driving forest degradation and fragmentation. Addressing the conflicts between rural livelihoods and forest restoration is paramount for long-term sustainable action. Box 11.3 illustrates how important providing access to adequate healthcare is to reversing forest fragmentation. These examples highlight that, even in communities where livelihoods remain insecure, a passion for the environment exists and can be fostered not only to restore habitat but also to link these local activities to help secure better incomes and protect ecosystem services.

These case studies provide examples of where people's passion and dedication have had a remarkable impact in reversing forest fragmentation at the local scale. With time, these are likely to have knock-on effects to increase forest restoration and enhance natural recovery of forest through increased availability of seed for natural forest succession, better management of remnant forest fragments, and conservation and sustainable management of the fauna that persist in these landscapes. Scaling up and accelerating the pace of establishment of smaller projects is essential if the current global targets are to be attained. Establishing links among different projects and establishing restoration projects on larger scales will facilitate forest recovery but requires major financial investment, and it is perhaps of little surprise that one of the largest restoration projects in the Asian tropics, the Harapan Rainforest (Box 11.4), is sponsored by one of the largest conservation organizations in Europe, the UK's Royal Society for the Protection of Birds. Securing the resources to prevent further degradation and planting of over a million trees on this 98,554 ha area of selectively logged, dry lowland rainforest in Sumatra is a significant challenge, which is not sustainable without significant long-term financial support.

Finally, if forest fragmentation is truly to be addressed on a scale that has significant impact globally, then resolving the conflicts between industrial agriculture and landscape-level planning are urgently required. Large-scale industrial agricultural activities such as palm-oil production or soya cultivation are major threats to both intact and

Box 11.1. The Carrifran wildwood project in Scotland

A few millennia after the end of the last Ice Age, the British Isles was almost entirely covered with natural forest collectively known as 'wildwood'. The following centuries saw most of this forest converted to alternative land use by grazing, felling and burning, so that today less than 5% of the original forest remains. The current UK government targets are to increase woodland cover to 12% by 2060 (DEFRA, 2013). In 1995, a group of residents in the Southern Uplands of Scotland set out to recreate a wildwood on the rolling hills, which had been almost entirely denuded of trees by past agricultural practice (Fig. 11.1). This was perhaps not your typical group of 45 local residents, as they included a number of professional ecologists, zoologists, botanists and foresters, together with other professionals with skills in law, education, IT and finance. What united this group of people was the passion to recreate a natural forest environment for all people and wildlife to enjoy. This was the start of the Carrifran Wildwood project (http://www.carrifran.org.uk/). Since its inception, the project has purchased almost an entire catchment area of 650 ha, the Carrifran Valley. Based almost entirely on volunteer support and charitable donations, this previously denuded landscape in just 14 years has been transformed. More than half a million trees have now been established at the site, with some 300 ha of the land now covered by native trees (Fig. 11.1). The project provides a flagship of community-based ecological restoration in the region and is an excellent example of the burgeoning 'rewilding' movement in Europe. Indeed, as of the end of November 2013, the Borders Forest Trust, the parent charity for Carrifran, purchased the Talla and Gameshope estates, an additional 1800 ha of denuded land immediately north of Carrifran, contiguous with it and with the National Trust for Scotland's property at Grey Mare's Tail, and very close to the other property of Corehead. This means that there is a real opportunity to restore something akin to the natural woodland and moorland ecosystem over more than 30 km^2 of the wildest part of the Southern Uplands. Thus, although Carrifran is relatively small, it has provided a stepping stone enabling a much more ambitious scale of restoration. This project has enhanced the capacity for the area to support biodiversity and reduced the fragmentation of the forest landscape by connecting up the relict trees in gullies and along streams. It also provides a showcase of how ecological restoration can be achieved and is a powerful education and awareness-raising resource, in addition to creating local employment in a rural area where agriculture has been in decline.

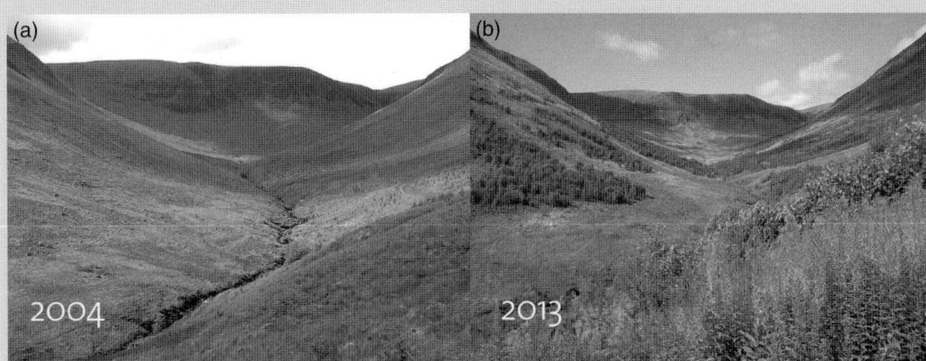

Fig. 11.1. Views of the Carrifran site at the start of planting in 2004 (a) and in 2013 (b). With kind permission of John Savory ©.

degraded forest in the humid tropics. Box 11.5 summarizes one way in which these conflicts might be reduced by combining carbon offsets with well-informed land-use decision making through a more holistic approach to industrial palm-oil production systems.

Conclusion

Reversing the trend in global forest fragmentation is essential, not only from an environmental perspective but economically and for the benefit of global society. Short-sighted

Box 11.2. Ecosystem restoration in Gunung Leuser National Park, Northern Sumatra, Indonesia

Around 4 million people living in the provinces of Aceh and North Sumatra depend on the Leuser forests for ecological services such as the provision of water for consumption and irrigation, soil fertility, flood control and climate regulation. The protection of this high-conservation-value forest should be integrated in the public agenda to maintain the sustainability of the critical ecological services of the Gunung Leuser National Park (GLNP), part of the Tropical Rainforest Heritage of Sumatra UNESCO World Heritage Site. This forest is considered the last stronghold of the remaining Sumatran orang-utan population. Efforts to halt the expansion of plantations in GLNP are urgently needed, including the restoration of forested areas that have already been illegally encroached upon. The Orangutan Information Centre (OIC) develops local initiatives to protect the park through the restoration of degraded land in the park buffer zone, and is managing a local community development project aimed at undoing the damage caused through the conversion of national park forest in Sei Betung, Besitang, into plantation agriculture by the management group PT Putri Hijau and PT Rapala, among other smaller offenders. This forest restoration aims to match (if not exceed) the original levels of species diversity, ecosystem structure and ecosystem function, while planting and encouraging natural regeneration of tree species that are known to have been present originally. In addition, the OIC aims to create a social environment that is conducive to and fully supportive of the continued conservation of the park, in order to better secure the future of the ecosystem and all of its inhabitants.

The OIC has so far been restoring the damaged areas in the Sei Betung resort of the Langkat district of North Sumatra, within the protected GLNP, where more than 500 ha of protected forests had been converted to oil-palm plantation by two relatively small companies. The programme objectives are as follows:

1. Restore areas of the GLNP in Sei Betung, Besitang, which have been degraded as a result of illegal agricultural encroachment, by replanting with indigenous tree species.
2. Conduct intensive maintenance on planted trees as well as naturally growing tree species, to best ensure their survival and growth.
3. Conduct species diversity enhancement planting on previously replanted areas.
4. Remove all remaining illegally planted oil-palm trees from the restoration site.
5. Monitor biodiversity levels in the restoration site.
6. Monitor the growth and development levels of planted trees.
7. Build the capacity of local stakeholders to become stewards of the replanting programme.

To date, 250 ha of national park forest have been replanted with close to 500,000 seedlings of 97 indigenous tree species. This work has been carried out with the full support of the local community, with a local farmers' group, KETAPEL (Farmer Association for Leuser Protection), formed through this programme, working in close collaboration with the field staff assistants. Each hectare of degraded land is planted with approximately 1100 seedlings of 20–30 different tree species, with an initial priority of planting fast-growing pioneer species in order to regrow the rainforest canopy. This is a vital aspect of rainforest ecology, providing shade for growing trees from extreme heat, with fallen leaves providing a richer soil in which they can take root. It is only with these conditions that seedlings can overtake the otherwise dominant grasses growing in the degraded area. The restoration of these natural processes enhances the forest's potential to regenerate naturally.

In 2012, the complete removal of all illegally planted oil-palm trees from the restoration site was finally achieved, through the removal of the last remaining 4800 oil-palm trees from 48 ha of formerly pristine GLNP forest. Other project developments have included new species sightings from our wildlife monitoring initiatives, including the first sightings of wild Sumatran orang-utans in the restoration area.

A daily climate monitoring initiative has also begun in the site, and new uniforms have been distributed to the OIC and KETAPEL restoration team. Furthermore, the number of community education and outreach activities in the programme has been increased. In addition to distributing new information materials to the local community, a number of local schools have visited the site, as well as other domestic and foreign visitors. Many of these visitors, including Indonesian celebrities and media, have also stayed overnight in local homestays, which has provided an additional source of income for local families. In addition, staff members have been visiting other restoration sites and exchanging knowledge with experts and forest-restoration organizations including the Forest Restoration Research Unit in Chiang Mai, Thailand.

Continued

Box 11.2. Continued.

Despite some minor setbacks, including drought and/or floods damaging planted trees, the project is going from strength to strength (Fig. 11.2). There is optimism about the future of this area of critical habitat, with the hope that every last hectare that has been lost will soon be on the way to being rainforest once again. Moreover, the project has brought stakeholders together working to rehabilitate degraded land within the park to manage the replanting of thousands of indigenous tree seedlings. Through this project, a significant shifting in the mind-set of the local community has occurred regarding conservation issues and the protection of the park, which is also part of the Tropical Rainforest Heritage of Sumatra UNESCO World Heritage Site. The programme has also resulted in more positive views about the need to sustain and protect the forests from encroachment, not only because the ecosystem has been granted full legal protection by the national government but also to sustain the valuable ecological services that forests freely provide, which are essential for the daily lives of both wildlife and people. Local people are participating in training and capacity-building exercises in support of the restoration work, including tree nursery development, improved planting methodology, planting maintenance and wildlife monitoring, and they also benefit financially through receiving a stipend for programme-related assistance. All of this, in addition to helping restore this degraded tract of forest, has already resulted in local communities realizing that they are enhancing the natural ecological services, including restoration of the local water table and drought resistance, previously lost due to forest clearing and monoculture agricultural development.

Source: Panut Hadisiswoyo, Founder and Director of the OIC, Medan, Indonesia.

Fig. 11.2. Before (a) and after (b) photographs of ecosystem restoration at Gunung Leuser National Park. © Panut Hadisiswoyo.

decisions where forests are valued below that of short-term economic gain will continue to erode biological diversity, compromise ecosystem functioning and services, and undermine human health and livelihoods. This book demonstrates that our knowledge and understanding of the implications of forest fragmentation are sufficient

Box 11.3. Stethoscopes and seedlings: aligning human and rainforest health

Human health and rainforest health have always been tightly coupled. Poor health in rural communities frequently leads to the degradation of nearby forests. For example, one of the main incentives for small-scale illegal logging in Indonesian Borneo is the need for cash to pay costs associated with healthcare emergencies. In turn, degrading forests can increase downstream flooding and droughts, and reduce agricultural yields and thus villagers' health. Degraded forests also have larger mosquito populations, which increases the risk of human disease. To directly address this vicious circle, conservation initiatives need to proceed hand in hand with healthcare activities. Here, we report the work of one such program, Alam Sehat Lestari (ASRI; http://alamsehatlestari.org), operating since 2007 in communities around the Gunung Palung National Park in West Kalimantan. ASRI offers local residents high-quality, affordable healthcare, directly integrated with a portfolio of forest-conservation activities.

The 90,000 ha Gunung Palung National Park is one of the most biodiverse parks in Indonesia, and the world, containing a nearly complete set of Borneo habitats, from mangrove and lowland alluvial forest to montane forest. The park is home to approximately 2500 wild orang-utans (*c.*10% of the world's surviving population), and an intact large-mammal community, including clouded leopards and sun bears. However, the park is now an island of forest in a sea of oil-palm plantations, and members of the surrounding human population are forced to meet their timber needs, and often their need for ready cash, by felling and hand extracting trees of the many valuable timber species in the park. Additionally, high housing and infrastructure growth in the region is creating a large demand for the timber.

ASRI's approach is to listen carefully to local communities and then offer sustained activities that meet the needs and aspirations of the community, while promoting a reduction in illegal logging. Many of the villagers are poor (mean household income was estimated at $13 per month in 2003), and medical emergencies often lead to major debts, as government hospitals are distant and expensive. Illegal logging is one of the few ways to earn a significant cash income, but the work is dangerous and few would actively choose it over legal alternatives. Locals also have a well-developed understanding of their dependence for water and food on intact rainforest. In response, ASRI, the flagship programme of US-based Health in Harmony (http://www.healthinharmony.org/), founded by a US physician (Kinari Webb) and Indonesian dentist (Hotlin Ompusunggu) and working closely with local government, has implemented the following:

- A central health clinic, providing high-quality, low-cost medical care to local villagers, who can also pay with non-cash items (e.g. manure, seedlings) and labour. Prices at the clinic are linked to the level of illegal activities in the patients' area, thus providing a direct incentive for conservation.
- A mobile clinic unit, which visits remote communities around the park each month.
- Sustainable agriculture training, assisting local farmers to increase their income while improving soil quality, and also demonstrating the drawbacks of 'slash-and-burn' farming. A related 'Goats for Widows' programme provides income for widows and goat manure for the farmers.
- A reforestation project, demonstrating forest-restoration methods in a degraded but accessible part of the park and employing local communities.
- A Forest Guardians programme, whereby a member of each of the 31 participating villages around the park acts as a conservation liaison partner, listening to community needs and spreading information about alternatives to illegal logging.

ASRI's approach is explicitly long term, in recognition that trust takes time to build yet is a fundamental component of conservation success. Repeated questionnaire surveys of health and economic indicators show that illegal logging has fallen since the programme began and community health has improved significantly. While ASRI activities are always part of a complex, dynamic social and economic landscape, the evidence points clearly to the effectiveness of this integrated healthcare and conservation model for reducing forest degradation rates.

Since 2009, the forest-restoration project has planted native species on 26 ha of logged, burned land within the park boundary at two sites: the first is an area in the south that was clearcut by a rogue logging firm in the 1980s, while the second suffered from illegal logging and agriculture but now provides an important 'orang-utan corridor' between two larger areas of forest. At both sites, a major challenge for reforestation is plant competition by rapidly growing ferns and grasses, especially the aggressive *Imperata cylindrica* ('alang-alang'). In order to establish a closed canopy as soon as possible to shade

Continued

Box 11.3. Continued.

out the competition, the primary intervention has been a dense, one-time planting of native tree species (up to 4250 seedlings ha^{-1}). The species composition has been a mix of pioneer and shade-tolerant species, and has surprisingly demonstrated that many of the top-performing species in full sun have been in the latter class, destined to become forest giants. The programme has also experimented with methods of accelerated natural regeneration, not planting but liberating existing saplings from their competition. As well as employing a permanent team to weed, measure sapling growth and stock the seedling nursery, the project employs hundreds of locals for shorter-term efforts such as site preparation and planting events.

Another major challenge has been fire prevention. Although the park receives over 4 m of rain a year, there can be extended periods with little precipitation, and the dried ferns and grasses burn easily. Despite the successful suppression of nearby fires by staff and community members in previous years, a fire began and raced through the 20 ha southern reforestation block in September 2013, killing most of the saplings planted over the previous 4 years. Tragically, such burning of restoration areas in the tropics is a relatively common occurrence. After the fire, both staff and community were unanimous in the desire to continue work in some way to demonstrate a commitment to the land, and 4 ha of the 20 were successfully replanted in December 2013. At ASRI, reflection on the fire has actively moved from feelings of loss of the new forest (a hoped-for 'end') to a recognition of the undiminished benefits of the way the work was conducted (the 'means'). For example, the desire to reforest the land was deeply shared by locals and outsiders alike, and the work cemented the relationship between all parties. Similarly, in the act of planting and tending, the reality of how hard it is to replant even 20 ha was made clear, and community members have expressed that they are more likely to think twice now before they cut down a tree. Thus, although the route to regreening these hills can been long and turbulent, the effort is never wasted.

Source: Campbell O. Webb (Arnold Arboretum of Harvard University) and Kinari E. Webb (Health in Harmony, and Yayasan Alam Sehat Lestari)

Box 11.4. Harapan Rainforest, Sumatra, Indonesia

Unsustainable logging of tropical forests has resulted in large areas of degraded forest throughout the tropics. Although these forests are often legally designated as forest estate, they may still be converted to plantations or agriculture when timber resources are exhausted. Moreover, once forests are no longer producing timber and operations are scaled back, they become vulnerable to further degradation through illegal logging, encroachment and fire, as a result of inadequate protection. Nevertheless, such forests occupy enormous swaths of land. In Indonesia alone, there are approximately 24 million ha of former logging concessions without current management, representing an enormous opportunity, but also a significant challenge, for conservation.

Harapan Rainforest is an ecosystem restoration concession covering 98,000 ha of lowland Sumatran rainforest and was the first such licence awarded. The concession license is held by a company (PT REKI), which is a partnership between Burung Indonesia and the Royal Society for the Protection of Birds, UK. The project has received substantial donor funding, particularly from the Danish International Development Agency (Denmark) and KfW (Germany). Preliminary surveys have demonstrated that the forest harbours a remarkably diverse fauna, with 302 bird species and many rare mammals, including tigers and elephants. Nevertheless, most of the forest is highly degraded, with over 80% of sample plots supporting <50% of the mean number of large trees (>30 cm diameter at breast height) found in nearby undisturbed forest. Large areas of the forest burned in the widespread fires in 1997/1998, and encroachment is a significant problem.

Restoration of tropical forests on this scale has not been attempted before, and there are significant technical hurdles. Currently, a number of treatments are being trialled, including selective thinning of pioneers and direct seeding. However, perhaps even more challenging are the social and economic issues. Donor funding has been essential in establishing Harapan Rainforest, but it is not a financial

Continued

Box 11.4. Continued.

model that is sustainable in the long term, nor one that can be rolled out on a large scale. Moreover, the legitimacy of forest restoration as an appropriate use of land is continually being challenged by encroacher communities and local politicians with an eye to increasing support among their constituency. For ecosystem restoration to succeed in Indonesia and elsewhere will require the development of enabling business and social models. At Harapan Rainforest, multi-use forest management is being developed that includes conservation, timber production, non-timber forest products, agroforestry and small areas dedicated to intensive production of high-value forest products. It is hoped that the income generated will provide the company with the means to manage the land effectively and at the same time support livelihoods for forest and forest edge communities. Harapan means 'hope' in Indonesian and the hopes of many are riding on it.

Source: Rhett D. Harrison, World Agroforestry Centre, China, and Royal Society for Protection of Birds, UK.

Box 11.5. Project POTICO (palm-oil, timber and carbon offsets), forest and landscapes in Indonesia

Resolving the conflict between demand for land for large-scale production of agricultural crops while maintaining forest cover is a major challenge in reversing global forest fragmentation. As one of the largest emitters of greenhouse gases from conversion of forest to alternative land, Indonesia plays a key role. This partner project between the World Resource Institute and the largest coated-paper manufacturer in the USA, NewPage Corporation, aims to develop novel mapping techniques and community engagement to prevent the further loss and degradation of forest from palm-oil expansion and timber exploitation in Indonesia (http://www.wri.org/our-work/project/forests-and-landscapes-indonesia). The principle objective is to ensure that palm-oil production is only established on degraded land and thus credibly satisfies criteria for emerging carbon-offset markets.

to demand its reversal and prevention. The number of initiatives, at local, regional and international levels, to implement this knowledge are growing in number and gaining considerable momentum. The question remains of whether these changes are taking place quickly enough. To have the required impact at the global scale, we need to see a sustained increase in the scale of action, where forest restoration and integration of multi-use forest management within production landscapes are mainstream rather than the exception. This will require the establishment of policy at all levels, from the local to the global, to ensure that future landscape planning places forests and forest resources at every level of the decision-making pyramid. We hope that the scientific knowledge presented here can help to inspire and inform people working in all aspects of land management to sustain and restore forest landscapes to provide a rich and fruitful habitat for generations to come.

References

Anon. (2013) World Population Prospects, The 2012 Revision. United Nations, New York <http://esa.un.org/wpp/> (accessed 29 July 2014).

Asner, G.P., Rudel, T.K., Aide, T.M., DeFries, R. and Emerson, R. (2009) A contemporary assessment of global humid tropical forest change. *Conservation Biology* 23, 1386–1395.

CBD (2011) *Aichi Biodiversity Targets.* Convention on Biological Diversity (CBD) <http://www.cbd.int/sp/targets/> (accessed 29 July 2014).

Corlett, R.T. (2013) Becoming Europe: Southeast Asia in the Anthropocene. *Elementa: Science of the Anthropocene* 1, 000016.

DEFRA (2013) *Government Forestry and Woodlands Policy Statement.* Forestry Commission, Department for Environment, Food and Rural Affairs, UK.

Harrison, R.D., Tan, S., Plotkin, J.B., Slik, F., Detto, M., Brenes, T., Itoh, A. and Davies, S.J. (2013) Consequences of defaunation for a tropical tree community. *Ecology Letters* 16, 687–694.

Holl, K.D. and Aide, T.M. (2011) When and where to actively restore ecosystems? *Forest Ecology and Management* 261, 1558–1563.

ICUN (2011) *Leaders Define Pathway to Restoring 150 Million Hectares of Lost Forests*. International Union for Conservation of Nature <http://www.iucn.org/?uNewsID=8147> (accessed 29 July 2014).

Ismail, S.A., Ghazoul, J., Ravikanth, G., Kushalappa, C.G., Uma Shaanker, R. and Kettle, C.J. (2014) Forest trees in human modified landscapes: ecological and genetic drivers of recruitment failure in *Dysoxylum malabaricum* (Meliaceae). *PloS ONE* 9, e89437.

Kettle, C.J. (2012) Seeding ecological restoration of tropical forests: priority setting under REDD+. *Biological Conservation* 154, 34–41.

Lamb, D. (2011) *Regreening the Bare Hills*. Springer, Dordrecht, the Netherlands.

Sayer, J., Sunderland, T., Ghazoul, J., Pfund, J.L., Sheil, D., Meijaard, E., Venter, M., Boedhihartono, A.K., Day, M., Garcia, C., van Oosten, C. and Buck, L.E. (2013) Ten principles for a landscape approach to reconciling agriculture, conservation, and other competing land uses. *Proceedings of the National Academy of Sciences USA* 110, 8349–8356.

Schmitt, C.B., Burgess, N.D., Coad, L., Belokurov, A., Besancon, C., Boisrobert, L., Campbell, A., Fish, L., Gliddon, D., Humphries, K., Kapos, V., Loucks, C. *et al.* (2009) Global analysis of the protection status of the world's forests. *Biological Conservation* 142, 2122–2130.

Stanturf, J., Lamb, D. and Madsen, P. (2012) *Forest Landscape Restoration Integrating Natural and Social Sciences*. Springer Science+Business Media, Dordrecht.

WRI (2011) *A World of Opportunity for Forest and Landscape Restoration*. World Resources Institute <http://www.wri.org/resources/maps/global-map-forest-landscape-restoration-opportunities> (accessed 29 July 2014).

WWF (2002) *Forest Management Outside Protected Areas*. World Wildlife Fund <http://awsassets.panda.org/downloads/po3aforestmanagementoutsidepas_m149.pdf> (accessed 29 July 2014).

Index

Page numbers in **bold** type refer to figures, tables and boxes.